Field and Forest

THE NATURALIST'S BOOKSHELF

The Naturalist's Bookshelf is a contemporary series of new and original works on environmental concerns, gardening, and nature study. Innovative in both design and content, volumes in this series are for your personal library reference and use in the field.

White House Landscapes: Horticultural Achievements of American Presidents
Barbara McEwan

The Naturalist's Path: Beginning the Study of Nature
Text and Illustrations by Cathy Johnson

The Nature Directory: A Guide to Environmental Organizations
Susan D. Lanier-Graham

The American Littoral Society Handbook for the Marine Naturalist
David K. Bulloch

The Pleasures of Watching Birds
Lola Oberman

Exploring Nature in Winter
Alan M. Cvancara

FIELD AND FOREST

A Guide to Native Landscapes for Gardeners and Naturalists

Text and Illustrations by Jane Scott

WALKER AND COMPANY
NEW YORK

To Tony
who helped me all the way

Copyright © 1984, 1992 by Jane H. Scott

All rights reserved. No part of this book may be reproduced or transmitted in any form or by any means, electronic or mechanical, including photocopying, recording, or by any information storage and retrieval system, without permission in writing from the Publisher.

First published in the United States of America in 1992 by Walker Publishing Company, Inc.

Published simultaneously in Canada by Thomas Allen & Son Canada, Limited, Markham, Ontario

Library of Congress Cataloging-in-Publication Data
Scott, Jane, 1931–
Field and forest : a guide to native landscapes for gardeners and naturalists / Jane Scott ; illustrations by Jane Scott.
p. cm.
(The Naturalist's Bookshelf)
Includes index.
ISBN 0-8027-1193-6 (cloth). — ISBN 0-8027-7379-6 (paper)
1. Botany—Ecology. 2. Plant communities. 3. Landscape ecology. 4. Native plant gardening. 5. Landscape gardening. 6. Native plants for cultivation—North America. I. Title. II. Series.
QK901.S34 1992
581.5—dc20   92-3478
CIP

Book design by Georg Brewer

Printed in the United States of America

1 2 3 4 5 6 7 8 9 10

· CONTENTS ·

Foreword vii
Acknowledgments xi

I
THE LANDSCAPE
1. The Community of Plants 3
2. Man and the Land 11
3. Plant Succession 18
4. Plant Provinces 25

II
PLANT COMMUNITIES
5. Deciduous Woods 37
6. Open Lands 52
7. Wetlands 72
8. Dry Lands 95

III
ECOLOGICAL LANDSCAPING
9. Your Garden as a Plant Community 113
10. Nature by Design 117
11. Maintaining the Natural Garden 126
12. Releasing the Native Landscape 133
13. Acquiring Native Plants 148

IV
THE PLANTS
14. What's in a Name? 157
15. Identifying Plants 168

Appendix A. Resources 173
Appendix B. Organizations 175
Bibliography 181
Index of Plant Names 185

· FOREWORD ·

I am frequently asked by visitors, captivated by the naturalistic gardens of Mt. Cuba Center here in Delaware, where they can find help in making such gardens for themselves. The books I can recommend are few, and most of them long out of print. Fortunately, this new work by Jane Scott is an easy-to-read, inspirational, and instructive introduction to gardening in a natural way. It is a book I can recommend without reservation to those who wish their gardens to be personalized extensions of the beauty they find in nature.

Field and Forest: A Guide to Native Landscapes for Gardeners and Naturalists explains how nature functions, how the landscape around us was formed through time by the interaction of all the species comprising our biotic communities, and how man has intentionally and unintentionally impacted these dynamic communities. After reading the book, we come away inspired and with heightened sensitivity to the natural world on which the spirit depends; and, for those of us who wish to go farther, with a basic understanding of how to capture the essence of natural landscapes in our gardens.

In the early chapters Jane Scott guides us skillfully through the complexities of plant communities, plant succession, and plant provinces, showing us, along the way, the measure of joy available to those who understand the

∴ FOREWORD ∴

processes on which rest the beauties of form and pattern. She brings us also to an understanding of how the human species, by many seemingly innocuous acts, has upset these ancient, evolved systems.

Later chapters just as skillfully describe an approach to gardening based on that understanding and appreciation of nature established in the earlier chapters. This is no tedious academic discourse but a joyful journey into the realm of creative gardening. Jane Scott helps us to grasp the principles of naturalistic gardening and, through the craft of gardening, to put these principles into practice. For those who delight in the diversity and complexity of the natural world, she discusses how to reclaim landscapes smothered by the exotic plant pests we have, in many cases, intentionally introduced. And for those who wish to be a part of the creative process of gardening, chapters on naturalistic design and maintenance provide the essential understanding needed to make and care for landscapes. This is not a book for those who want recipes; it is a work for those who wish to understand.

Throughout the book Jane Scott's conservative ethic provides an underlying theme, and this is nowhere more obvious than in the chapter on purchasing plants. She teaches us how to spot plants that have most likely been dug from wild populations and, most important, why it is necessary to curb our desire to grow some of our most beautiful and threatened natives.

In short, this is a book whose time has come. It is packed with information useful to the new generation of thoughtful nature lovers, whether they garden or not. It is concise and enjoyable to read and abundantly illustrated with the drawings for which Jane Scott is well known. It is, finally, a book that addresses the processes of nature and gardening, and leaves the form of the product to each of us.

FOREWORD

This creative process is, after all, where much of any gardener's personal joy and satisfaction must reside.

—Richard W. Lighty, director,
Mt. Cuba Center for the
Study of Piedmont Flora,
Greenville, Delaware

· ACKNOWLEDGMENTS ·

The knowledge and enthusiasm of innumerable botanists and gardeners, past and present, provided help with this book. Although it is impossible to list them all, special thanks go to Dr. Richard W. Lighty, director of the Mt. Cuba Center in Greenville, Delaware, for his patient help and advice, and to Dr. Donald Huttleston, former taxonomist at Longwood Gardens, Kennett Square, Pennsylvania, for sharing his knowledge generally and especially on trips to the Great Smokies and the New Jersey Pine Barrens.

Leslie Sauer, of Andropogon, Inc., and F M Mooberry are among the many native plant enthusiasts who helped with the sections on gardening and restoring the native landscape. I am also indebted to George Fenwick, who introduced me to a freshwater marsh on Maryland's Eastern Shore, and to the late Hal Bruce who led me through the Winterthur meadow. Norman H. Dill and Arthur O. Tucker, both of the Department of Agriculture and Natural Resources at Delaware State College, took me to see rare plant communities on the Coastal Plain. John and Mary Watkins helped to explore the oak woods of the Shenandoah and also contributed material on the pine barren communities of North Carolina. Ann R. Daudon, Nancy Frederick, and Elizabeth Sharp also provided valuable help and support. I am indebted to them all.

I
The Landscape

·O·N·E·

The Community of Plants

SKUNK CABBAGE
Symplocarpus foetidus

This is a book for people who love plants. It is for those who hunt spring wildflowers in the woods and summer ones in meadows, for those who wander along beaches and wade into marshes, and, above all, for those who long for a garden that reflects the authentic beauty of the natural landscape. You will not, however, find a description of every plant you see. This is not a field guide, but an attempt to put plants into context.

Putting a plant into *context* means to show it as a member of a community, that is, a group of species that have adapted to the same habitat. Some plants grow in the moist soil of woodlands, others hug the driest seaside dune, and still others push up through cracks in a sidewalk. If you do not need to be told that the same species will not do all three, you already know something about plant communities.

There will always be plants in the landscape; they are among the most persistent of life forms. Yet some plants are very set in their ways. Having adapted to the requirements of their habitat over millions of years, they are destroyed by the slightest disturbance. Others are flagrant opportunists, putting down roots wherever they get the chance. The numerical relationship of one kind to the other tells us enlightening, but often painful, tales about our past use of the land. We learn to read the landscape like a book.

The more chapters we read of this "book," the more

we come to value the original native plant communities, not only because they are the product of millions of years of evolution, but also because they clothe the land with an orderliness and beauty that enhances our sense of place — that elusive quality that is so often erased by commercial mass production.

Of course, modern civilization requires some disturbance of the land. Yet in our ignorance, we destroy far more than we need to, often making it impossible for the land to heal. It does not have to be this way. The more we know about the native plant communities that surround us, the more we will come to appreciate their inherent beauty and diversity and the more effectively we can accommodate them in the places where we live.

Among the special appeals of botany is the fact that plants, unlike animals, do not vanish at our approach. Yet they do move in time, shifting images with the revolving seasons like a slow-motion kaleidoscope, as bright fruits replace delicate flowers and the pervasive green of summer explodes into autumn's fire. In winter, every species of tree shows a distinctive silhouette, enhanced perhaps by caps of snow or ice gleaming in the sun, and down in the meadow we can inspect the seed heads of grass and goldenrod and speculate on the promise of each round green rosette that hugs the frozen soil.

As spring progresses, we may find that certain wildflowers have disappeared from well-remembered sites and reappeared somewhere else. Individual plants cannot move, of course, but whole populations seem able to migrate in mystifying leaps, sometimes over long distances, thanks to that minute miracle: the seed.

Seeds are now so commonplace, it is hard to imagine just how important their development was to the evolution of plants. The ferns, horsetails, and club mosses of today still reproduce by spores in the primitive two-stage manner of the first land plants of hundreds of millions of years ago. The

∴ THE COMMUNITY OF PLANTS ∴

mature plant releases spores from tiny capsules to blow on the wind. If they land on moist ground, they develop into the second stage of reproduction: a tiny plant called the prothallus containing the male and female organs. Like the sea plants from which it evolved, the prothallus has swimming sperm and is dependent on the presence of water, in this case merely a film, to accomplish successful fertilization.

The conifers were the first plants to have true seeds. They also had male and female cones on the same tree, eliminating the uncertainties of an intermediate generation. Each cone has scales arranged spirally around a central axis. Pollen from the small male cones is dispersed by the wind. When it successfully pollinates the ovules in the larger female cones, seeds develop behind the scales. That this has proved to be an efficient method of reproduction is demonstrated by the abundance of conifers still found on the earth, yet the opportunities for pollination and seed dispersal pale beside those of the next evolutionary group: flowering plants.

There were no insects on earth when the conifers first evolved, but by the time flowering plants had appeared, one hundred and thirty million years ago, they were plentiful. Conspicuous flowers are, in fact, an opportunistic attempt to entice and employ insects for purposes of pollination. As time went on, both insects and flowers became more and more specialized and adapted to each other's needs. Insects were able to carry the pollen from flower to flower with a precision not possible when fertilization depended on the capriciousness of the wind. The seeds, too, became more specialized. Now they developed within the protection of an ovary and gradually became equipped with varied and ingenious methods of travel. Today there are seeds that fly on the wind with tufts of hair for parachutes, and others, like the seeds of maples and ashes, that are shaped like propeller blades and twirl through the air like tiny helicopters. There are seeds so fine they blow about like dust, and oth-

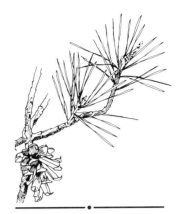

VIRGINIA PINE
Pinus virginiana

ers that hitch rides on the feet of wading birds. Delectable fruits often have indigestible seeds, a subtle but effective means of travel, whereas stick-tights and burrs simply cling to the coats of animals or humans.

Because all these seeds are sprinkled randomly over the landscape, one would expect a random mixture of species to result, yet even the most cursory glance tells us this is not the case. Woods do not mix with marsh, and the plants found in a grassy meadow are not the same as those growing on a mountain slope.

The truth is that plants are cliquish. Similar species occur in similar groups in similar habitats all over the globe, whether these habitats are contiguous or not. Each plant in a community arrives there independently, but rarely does a single species occupy a site to the exclusion of all others, with the possible exceptions of the red mangrove, *Rhizophora mangle*, in coastal Florida and the stands of giant reed, *Phragmites australis*, that have spread over disturbed marshes in the Northeast.

Habitat, then, is the key that determines plant communities, but what determines a habitat? Over a large geographical area, the answer is climate. Yet this is only part of the story, because numerous and diverse minihabitats occur within broad areas of similar climate, depending on the available sunlight and water as well as the general character of the soil. In fact, several minihabitats may occur on a single hillside causing different plant species to occupy the top, the bottom, and the northern and southern slopes.

As in human communities, individual species in a plant community both support and compete with one another. Nor is the habitat the same for all. The crown of a forest tree basks in bright sunlight, whereas the ferns at its base grow in dense shade. Yet without the tree to shelter them and its decaying leaves to provide humus for the soil, the woodland ferns could not grow. Thus, it is the increasing shade of a developing woods that makes it possible for

∴ THE COMMUNITY OF PLANTS ∴

new species to migrate there; in other words, the plants themselves gradually alter their own habitat.

In botanical terminology, a *stand* is a plant community that is homogeneous in all its layers, and an *ecotone* is the boundary between differing stands. An ecotone may be as abrupt as a cliff or as gradual as a sloping hillside, depending on whether the plant species that dominate one community mingle with those of another before disappearing. Ecotones may exist in time as well, as one set of plant species is replaced by a succeeding group on the same site.

When botanists speak of the range of a plant, they mean the whole geographical area over which it is found. In field guides, the phrase "from Labrador to Georgia" usually means a plant that is widespread in the north but is found only high in the mountains farther south. Conversely, "Florida to New Jersey" means a southern species, abundant in the lowlands of the South that because of the moderating effect of the Gulf Stream has been able to migrate northward along the Coastal Plain. Ranges can be discontinuous, sometimes startlingly so. For instance, many genera of familiar North American plants, such as rhododendrons, witch hazel, sassafras, and magnolias, are also found in Asia. The genus of the tulip poplar, *Liriodendron tulipifera*, contains only two species in the entire world, one in the eastern United States and the other in eastern China! Botanists think both populations are survivors of a vast temperate forest that covered the earth millions of years ago.

The rarity or commonness of a particular species refers to the distribution throughout its range. A plant that is thinly spread over a large range may not be classified as rare even though it is never plentiful at any one location, whereas a rare plant may be locally abundant but have a severely limited range. Plants become rare or even extinct for a variety of reasons. Most common, and most disturbing, is the incessant destruction of habitat. Man has always destroyed habitat to improve his own condition, but now,

GOLDENSEAL
Hydrastis canadensis

on the eve of the twenty-first century, both his numbers and his mechanical efficiency have increased to such a degree that a plant or animal disappears somewhere on the earth almost every day. Some of this is natural and beyond our power to stop, but all too much of it is due to man's recklessness and ignorance. Gardeners, in particular, should be aware that overcollecting, especially by commercial growers, has threatened the survival of certain plants. Others, such as ginseng, *Panax quinquefolius*, and goldenseal, *Hydrastis canadensis*, have been stripped from the woods because of alleged medicinal qualities. We are destroying valuable genetic diversity, which may indeed include the loss of potential medicines and food, in addition to diminishing the quality of life on this planet.

Particularly vulnerable are habitats with unusual and specialized growing conditions. The plants they contain have evolved specifically to fit those conditions and are ill-equipped to survive in other environments. Change the conditions and you inevitably lose the plants.

On the other hand, certain plants that were once rare may now be common because the habitats they prefer have become widespread. Many ubiquitous garden weeds fall into this category. They are plants that originally evolved in areas of severe disturbance. Before the spread of mankind, such places were limited to lands subject to frequent natural disasters, such as floods and avalanches. Now that disturbed land is the norm, these once rare plants have so quickly taken advantage of the prevailing situation that we now call them weeds. At the same time, many of our original native species have declined to the point of rarity.

Some plants, however, may be rare because they have not yet migrated to the full extent of their range. Perhaps their methods of seed dispersal are inefficient, or stretches of inhospitable habitat too large to migrate across divide their present locations from other suitable environments. This may be one explanation as to why many plants can be successfully grown in gardens far to the north or south of their present geographical homes.

We know that plants tend to migrate from the center to the perimeter of their range and that the farther they travel from home base, so to speak, the less tolerant they become of changing environmental conditions. Even so, the range of a southern plant may gradually creep northward during a period of mild winters, or fingers of forest may invade the prairie over a decade of wet summers, but eventually a fierce winter or a brutal drought will stop them. Thus, it is the extreme rather than the average weather conditions that build the invisible "environmental fence" that marks the limit of a plant's range.

Of course, extremely cold winters or dry summers can affect the vigor of plants in the center of their range as well and may reduce both the number of species found in a community and the health of individual plants. Large trees are rarely killed by one or two bad years, although their growth may be slowed considerably. Perennials, too, will usually survive, even though affected in height and vigor. Annuals and biennials, on the other hand, may disappear completely, unless their seeds have enough long-term viability to survive a full year without germinating.

For a plant population to maintain its numbers, one seedling from each member must survive to reach maturity. That sounds easy, considering the huge number of seeds that are produced every year; yet a large proportion never manage to germinate at all (they are eaten by wildlife or fall on inhospitable ground) and the majority of those that do usually succumb within the first year.

Experiments in Louisiana's pitcher plant bogs have shown that natural hybridizing between species of pitcher plants tends to increase when the plants are under environmental stress, and this may well be true of other species as well. However, while hybridizing is an important means by which plants evolve, to succeed, the new species must be

BLOODROOT
Sanguinaria canadensis

able to reproduce true to type and to compete successfully with its own parents. This does not happen often. When it does, the new plant may also be able to compete outside its original habitat, thus enlarging the range of the genus. Plants also mutate often, but again the new plant must be able to compete and to reproduce in kind if it is to survive. Two notable examples of spontaneous mutations which man has reproduced for his own use are the pink dogwood, *Cornus florida*, and the navel orange. The seeds of a pink dogwood may or may not produce pink offspring, and the navel orange has no seeds at all. Thus both plants must be reproduced vegetatively from cuttings.

Over the millennia, plants have adapted to their ranges in fascinating ways. For instance, the seeds of many northern plants will not sprout until they have gone through a period of freezing weather, thus automatically ensuring the young seedlings the benefit of a full growing season before they must endure a northern winter. (Incidentally, the same mechanism prevents northern plants from extending their range into the South.)

Herbaceous plants of the temperate zone die to the ground in winter, a device that protects them from cold, breakage, and excessive drying. Deciduous plants go dormant, dropping their leaves to prevent transpiration while the ground is frozen and water unavailable. Tree bark, which is made largely of cork, also protects trees from winter's drying winds. The needles of northern conifers are far more resistant to cold and drought than the leaves of deciduous trees, and because they remain on the trees all year, are able to carry on photosynthesis over a much longer period. Some continue to do so until the temperature drops several degrees below freezing, making them particularly well suited to high altitudes and northern latitudes, where the length of the growing season is too brief for hardwood trees. Conifers also thrive in the Pacific Northwest, where 90 percent of the rainfall comes in the winter, leaving the summers too dry for deciduous trees.

· T · W · O ·

Man and the Land

FRANKLINIA
Franklinia alatamaha

We usually assume that when the Europeans first arrived on this continent, a vast, unbroken forest covered all the land from the Atlantic coast west to the Mississippi River; a forest full of ancient trees with massive trunks and foliage that soared a hundred feet above a carpet of lush ferns and wildflowers. Such a forest certainly did exist; there are records of black walnuts six feet in diameter and hollow sycamores big enough to use as a barn or even shelter a frontier family until their cabin could be built. However, some historians claim that this forest could not have been as widespread or as untouched as we suppose, because the indigenous populations had been burning it for centuries.

Fire was their principal tool for controlling their environment. They burned the woods to clear trees for plots of maize and squash; they burned the underbrush to deny cover for their enemies and to ease their own travel from village to village; but mostly they used fire to increase the amount of game. It is well known that deep woods harbor few game animals. Deer, especially, need the cover and browse of young growth, and most of the plants that produce edible berries and seeds need sunlight in order to bloom. So the native peoples burned large areas of forest to produce ecotones of second growth where these plants could flourish and multiply.

Contrary to popular impression, controlled burning is

not necessarily destructive to woods. Small fires can even increase soil fertility while preventing the accumulation of excess litter on the forest floor that could eventually fuel a major conflagration. Yet it would be naive to assume that these early people were always in control of their fires, especially when fire was used as a weapon of war.

Fire does alter the distribution of species, favoring some plants over others. For instance, while oaks and pines are known to be more common in a woods that has been frequently burned, there is also intriguing evidence that trees such as the rare southern yellowwood, *Cladrastis lutea*, may also be fire dependent. This is not to say we should deliberately burn the woods to increase the incidence of such species, but only that fire plays a part in the development of an ecosystem in ways that are not yet fully understood. (Yellowwood, incidentally, is the single North American species of a genus also found in China and Japan. Its fragrant, white, pendulous flowers make it a wonderful tree for the garden.)

YELLOWWOOD
Cladrastis lutea

The Europeans were afraid of the forest, faced, as they were, with the need to sustain themselves on a vast empty continent without the support of a settled civilization. They, too, burned the woods to clear space for settlements and farms. What they didn't burn, they cut down.

England, as the mother country, believed its colonies had the duty to supply whatever raw materials it required, and what England principally required at that time was lumber, particularly oak timbers and pine masts for the Royal Navy. That its own forests had been depleted long ago and its admiralty forced to import wood from the Baltic coast did not, unfortunately, inspire the British to practice conservation in the New World. The crown sent emissaries deep into the forests of New England to mark the best trees by hammering an iron brand in the shape of a crown and anchor into the bark. This mark may still be seen on an occasional Maine giant that managed to escape the ax.

∴ MAN AND THE LAND ∴

By the eighteenth century, the colonists were building sturdy, often elegant houses as well as creating beautiful furniture from the abundant supply of wood. The cutting of the forest continued throughout the Revolution and beyond, and, as the lumbermen moved ever westward in search of virgin territory, they became folk heroes of Paul Bunyan-like proportions. Everywhere, the new settlers viewed the woods as their enemy. The trees not only interfered with farming, they also provided protection and cover for hostile attackers.

It must have seemed as if no one cared about the destruction of the woods, yet this was also a period of meticulous and enthusiastic interest in natural history. At that time, plants were the principal source of medicine, and the people of Europe were eager to find cures for such widespread complaints as "wasting fever," syphilis, and "pox." Take the example of sassafras, *Sassafras albidum*. The plant is familiar to us both as an attractive small tree of old fields and fence rows, and, where its suckering habit can be accommodated or controlled, as an interesting choice for the garden. Sassafras was singled out as a medicine tree as early as 1574 by Nicholas Monardes of Seville. In a pamphlet entitled "Joyful Newes out of the Newe Founde World," he recommended it as a remedy for malaria as well as for a variety of other ailments. The belief grew all over Europe that sassafras was a magic cure, and it became as valuable a commodity to the early settlers of Virginia as tobacco. Whole expeditions were sent from England for the express purpose of gathering the roots.

As time went on, many other American plants also became sought after, among them, blue lobelia, *Lobelia siphilitica*, as a cure for syphilis and Seneca snakeroot, *Polygala senega*, as a protection against snakebite. It wasn't long before a brisk traffic in seeds and plants was crossing the North Atlantic. In 1678, the Temple House Botany Club, together with the bishop of London (an enthusiastic bota-

BLUE LOBELIA
Lobelia siphilitica

SPRING BEAUTY
Claytonia virginica

nist himself, named Henry Compton), sent a young man named John Banister to collect plants in Virginia. By the early eighteenth century, two native-born Virginians, John Mitchell and John Clayton, were also describing the local flora. Their names are immortalized in two of the plants they found, *Mitchella repens*, the partridge berry, and *Claytonia virginica*, the spring beauty. By 1733 Mark Catesby had finished his beautiful illustrated classic, *A Natural History of Carolina, Florida and the Bahama Islands*, and farther north John Bartram had begun a collection of plants on a small farm outside Philadelphia. Born in 1699, John Bartram had an intense lifelong interest in botany. Although his formal education was slight, as a young man he had bought himself a botany text and learned Latin so he could read it. His first box of American seeds and plants was sent to London in 1734 to a Quaker merchant named Peter Collinson with whom he was to correspond for the next thirty years. In 1736 Collinson formed a group of interested botanists, each of whom paid five guineas in return for the seeds of a hundred North American trees and shrubs, each minutely described by Bartram. Many of these would become prized ornaments in English gardens. By midcentury, Bartram was respected in the highest levels of European society and in 1765 even became official botanist to the king of England. He continued to make trips into the American wilderness, traveling on a horse laden with leather bags full of his precious specimens. It was during a collecting trip to Georgia in 1770 that Bartram discovered a lovely small tree along the banks of the Altamaha River which he named *Franklinia alatamaha*, in honor of Benjamin Franklin. This tree was never found in the wild again after 1803. The cuttings taken by John Bartram, and later by his son, William, are widely credited with saving the Franklinia from extinction. In fact, the tree is now grown so successfully in gardens far to the north of Georgia, that some ecologists theorize that it was originally a northern species that was forced into a less than hospitable southern habitat by the glaciers.

William Bartram inherited his father's love for plants and combined it with a natural talent for writing. The colorful descriptions in his book *Travels*, published in 1791 after a four-year and five-thousand-mile trip through the southeastern United States, are credited with inspiring Samuel Coleridge to write his famous poem "Kubla Khan."

All these early botanists were deeply indebted to the great Swedish naturalist Carl Linnaeus. Linnaeus was born Karl von Linne in South Rashult, Sweden. His father, a parish curate, wanted him to study for the ministry, but his son's interest in plants was so intense that the senior von Linne was finally persuaded to send him to medical school, for at that time a knowledge of botany was the principal requirement for the practice of medicine. While in medical school, young Karl started a botanical garden, writing meticulous descriptions of all the plants. Because he wrote in Latin, he became known as Carolus Linnaeus. His new system for the scientific classification of plants, first published in *Species Plantarum* in 1753, was based on sexual characteristics. The number of stamens and pistils, the plant's male and female organs, determined the class and order to which it belonged (see Chapter 14). In addition, each plant was assigned one generic and one specific name, thus doing away with the long and complicated descriptions that had been previously used. Linnaeus's system was both simpler and more coherent than any that had preceded it, making it possible for botanists to describe new plants accurately and also to communicate clearly with one another for the first time in history. Eventually, as knowledge of biology grew, Linnaeus's system of classifying plants became constricting and was modified by later botanists, yet the simplicity and elegance of his binomial system of nomenclature continue to serve us well.

Linnaeus himself was deeply interested in the flora and

fauna of the New World and had sent his own envoy to America in 1748. This was Peter Kalm, the man for whom *Kalmia*, or laurel, is named. His book *Travels in North America*, also published in 1753, was widely sought after in a Europe long devoid of wilderness, and where the populated land had been encumbered by feudal restrictions for centuries. The Europeans thirsted for news of this vast new continent where individual property could be had for the taking, and strange new varieties of plants and animals were everywhere.

The European nations never recognized any political claim of the indigenous population to land in America, but immediately took title for themselves upon discovery and settlement. The English crown commonly granted charters from afar to colonial governors and organizations such as the Massachusetts Bay Company, which then distributed land to individuals in an effort to encourage further settlement. Thus, by the early eighteenth century most of British America was populated by independent farmers living on a few hundred acres cleared from the forest.

When the Revolution was over and the new Constitution ratified, all the land west of the Appalachians that had been previously claimed by the states was turned over to the federal government. It was then sold to individuals to help pay off the war debt, a practice that encouraged the surge of agriculture that was to sweep across the country over the next hundred years. More and more forests fell to ax and fire, and the wild prairie sod was severed by the plow. By the midnineteenth century, when the Industrial Revolution caused much of the population to move from farms to urban areas, even more trees were felled for charcoal to fuel the smelting furnaces of the burgeoning iron industry.

The remaining midwestern farms grew larger and their crops became geared to commercial grain markets, while the East shifted to the raising of sheep and dairy herds. In 1909, when James Taylor, headmaster of a school in Vermont, conceived the idea of the Long Trail, a footpath that still stretches from Williamstown, Massachusetts over the crest of the Green Mountains to the Canadian border, the slopes of many of those mountains were sheep pastures, grassy and bare of trees. Even today, it is possible to find the remains of stone walls that once fenced in the flocks near the crests of the ridges.

Today, the Atlantic Coastal Plain is still heavily farmed, but the hills of New England and the Piedmont proved too steep for efficient agriculture, and farmers found they could not compete with the crops flooding in from the Midwest or with the vast herds of the western range. Many of these farmers moved west themselves, while others simply gave up their fields to succession. Today those fields, many well into second growth forests, are being cleared once again for housing developments and shopping malls.

Many more eastern farms were abandoned during the Depression of the 1930s, when farmers could not pay off their debts. Near my house in southeastern Pennsylvania, there is a stand of tulip poplar, *Liriodendron tulipifera*, a species that acts as a pioneer hardwood in this area. Still visible on the floor of the woods are faint but regular corrugations: the remains of the last furrows left by the plow. Because that last crop was never planted, an immediate change took place in the field's environment, allowing the process of plant succession to begin, and the forest to return.

·T·H·R·E·E·

Plant Succession

DAISY FLEABANE
Erigeron annuus

According to the theory of plant succession, a field of plowed earth will, given seeds and time, eventually be transformed into a towering forest. The complex orchestration of species, each stage in its turn preparing the way for the next, begins the moment a piece of land is released from human domination. This rhythm of change is universal even though the players, or the particular selection of species, vary according to the location, site, and general characteristics of the soil.

First to appear are the annuals: those plants that germinate, flower, set seed, and die within a single growing season. Most of the weeds that are quick to cover bare ground and interfere with crops and gardens are annuals. They are also *ruderals*, or plants adapted, by means of rapid growth and seed production, to thrive on disturbed soil, an attribute that has perfectly equipped them to follow in the wake of man.

Typical ruderals include the quick-developing crabgrass, *Digitaria sanguinalis*, and purslane, *Portulaca oleracea*, that are able to clothe the bare dirt of an abandoned field almost immediately. They are soon joined by rosettes of biennials like Queen Anne's lace, *Daucus carota*, and mullein, *Verbascum thapsus*. (Biennials develop over two growing seasons; the rosette and root develop the first summer and the flowering stalk the second. The plant then dies after setting seed.) By fall, the rosettes of winter annuals like daisy fleabane, *Erigeron annuus*, and members of the mustard family, genus *Brassica*, have appeared as well. Winter annuals are typically plants of shallow sandy soils in areas of frequent summer droughts. Their seeds are able to delay germination until the cool wet weather of fall. They then pass the winter as a ground-hugging rosette and flower the following summer.

Strong-growing summer annuals like ragweed, *Ambrosia* spp., will also bloom the second summer after abandonment, along with fast-growing perennials like dandelion, *Taraxacum officinale*, and black-eyed Susan, *Rudbeckia hirta*. By the third growing season, however, these early plants will begin to give way to slower-growing perennials like goldenrod, *Solidago* spp., and asters, *Aster* spp.; scattered among them will be numerous small seedlings of the woody shrubs and trees that will dominate the next stage of succession.

It is at this point that the trouble begins. Nowadays, in many fields throughout the East, the majority of these woody plants are not natives but invasive aliens. They are imported shrubs and vines, here unimpeded by customary controls, whose rapid growth and efficient reproductive systems enable them to crowd out or entangle the native species, sometimes to such an extent that the normal pattern of succession is brought to a standstill. The fact that many of these plants are still sold as garden ornamentals in many parts of the country is a troubling indictment of growers and gardeners alike.

Among these garden escapees are rampant vines like Japanese honeysuckle, *Lonicera japonica*, and Oriental bittersweet, *Celastrus orbiculatus*, that twine over all other vegetation, choking out woody plants and smothering herbaceous ones. Japanese honeysuckle was introduced into the United States as a garden vine in 1806. Gardeners welcomed it for its fragrant flowers, as well as for the fact that it stays green for much of the winter, spreads easily, and climbs well.

FIELD MUSTARD
Brassica rapa

The last two characteristics, of course, soon transformed it into a serious nuisance in our woods and fields.

Oriental bittersweet was introduced as an ornamental vine from eastern Asia in 1860. As late as the 1960s, the seed was still being distributed to commercial growers by the National Arboretum. By that time, of course, it had also escaped into the countryside and now strangles native trees and shrubs from northern Georgia to central Maine. The native American bittersweet, *C. scandens*, is now rarely seen; it is thought to have hybridized with and been overwhelmed by the invader.

In some areas of the Northeast, another garden escape called Porcelain berry vine, *Ampelopsis brevipedunculata*, smothers young trees and shrubs. A relative of the grape, Porcelain berry does not twine, but lies on top of other plants like a blanket, robbing them of necessary light. Meanwhile, two imported shade trees, the Norway and sycamore maple, *Acer platanoides* and *A. pseudoplatanus*, have invaded many old growth woodlands, and an escaped garden perennial called purple loosestrife, *Lythrum salicaria*, has overwhelmed the natural vegetation of countless freshwater wetlands.

PURPLE LOOSESTRIFE
Lythrum salicaria

The common privet, *Ligustrum vulgare*, is another garden plant that now crowds into hedgerows and fields all over the Northeast, while its cousin the Chinese privet, *L. sinenese*, has become the dominant shrub in some areas of Louisiana. In Florida, an ornamental called Brazilian pepper, *Schinus terebinthifolius*, thrives on disturbed areas, while dense groves of bush honeysuckles like *Lonicera maacki* and *L. tatarica* are shading out native groundcovers from Pennsylvania to Massachusetts.

Finally, there is the multiflora rose, *Rosa multiflora*, a plant traditionally used as the rootstock for grafting hybrid tea roses until an enterprising nurseryman advertised it as a "living fence," some twenty-five years ago. Promoted everywhere as a barrier so dense and prickly that it would contain pets and livestock, multiflora rose was soon being planted along highways as well as in backyards and along fencerows. It proved so attractive to birds that it is now credited with the northern spread of cardinals and mockingbirds. They, in return, have spread it into nearly every open field in the East.

All these aliens are such strong competitors that they quickly transform an open field into a dense thicket of brush and entangling vines that leaves little or no room for native roses or blackberries, *Rubus* spp., sumac, *Rhus* spp., or viburnum, *Viburnum* spp. If they also succeed in strangling young tree saplings, the natural process of succession is virtually brought to a halt.

Of course, gardeners are not the only villains of the piece. Japanese knotweed, *Polygonum cuspidatum*, now found along railroad tracks and highways all along the Eastern Seaboard, is a member of the buckwheat family, Polygonaceae, and was originally sold to farmers as a quick cash crop during the Depression. Highway departments and soil conservationists are responsible for the spread of autumn olive, *Elaegnus umbellata*, planted en masse in the 1960s and now found in dense populations of up to fourteen thousand plants per acre in eastcentral Illinois. Garlic mustard, *Alliaria petiolata*, is an alien weed that is rapidly replacing native wildflowers in river bottoms, particularly in Illinois but spreading to other states as well, while the melaleuca or punk tree, *Melaleuca quinquenervia*, which was originally introduced into Florida to "forest" the Everglades, now threatens that entire ecosystem. Finally, there is kudzu, *Pueraria hirsuta*. First planted by the U.S. Department of Agriculture to control erosion, it now smothers millions of acres in the South and has crept as far north as Massachusetts.

In the Northeast, there is evidence that the problem is worse in fields that are underlain with metamorphic rock. The fact that the land was once farmed may also be a factor.

MULTIFLORA ROSE
Rosa multiflora

Farm fertilizers that raise the level of nutrients like phosphorus and nitrogen in the soil act to stimulate the growth of these fast-growing aliens and strengthen their grip on native vegetation.

The problem does seem to be far less severe on fields where the soil is poor, particularly if broomsedge, *Andropogon virginicus*, moves in first. Broomsedge is a native tuft grass with densely fibrous roots that spreads in dense swathes, interrupted only by small patches of goldenrod and the occasional red cedar, *Juniperus virginiana*, hawthorne, *Cretaegus* spp., or dogwood, *Cornus florida*. It was once thought that the soil in such fields was simply too thin to support other vegetation, but recent research has shown that *Andropogon* is also *allelopathic*. That is, its roots exude a chemical that inhibits other plant growth, a fact that could be put to good use by gardeners who are battling invasive aliens. Once established, fields of broomsedge seem able to persist for decades.

Two other grasses found on open ground and dry hillsides in the East are Elliott's beardgrass, *A. elliottii*, and little bluestem, formerly *A. scoparius* but now called *Schizachyrium scoparium*. Like broomsedge they turn a lovely red-brown color in winter.

Only when early pioneer trees such as hawthorne, dogwood, red cedar, *Juniperus virginiana*, sassafras, *Sassafras albidum*, and wild cherry, *Prunus* spp., are able to break through and become tall enough to shade the field in summer will these weedy plants weaken. Then, if the wind-driven seeds of tulip poplar, red maple, and white ash, *Fraxinus americana*, succeed in finding a foothold, they will grow quickly into tall stately trees that will eventually shade out the earlier hawthornes, cedars, and cherries. Young dogwoods now take their place as a forest understory tree, and shrubs such as mapleleaf viburnum, *Viburnum acerifolium*, spicebush, *Lindera benzoin*, and witch hazel, *Hamamelis virginiana*, begin to appear.

BROOMSEDGE
Andropogon virginicus

∴ PLANT SUCCESSION ∴

When squirrels entering the young woods bury nuts in the forest litter, small seedlings of beech, *Fagus grandiflora*, oaks, *Quercus* spp., and hickory, *Carya* spp., will sprout on the woods floor. Ferns will appear, followed closely by all the delicate wildflowers known as "spring ephemerals."

There is evidence that the deep shade and thick blanket of litter of the developing forest inhibits the germination of the light, airborne seeds of maples and tulip poplars, while favoring the heavier nuts of these forest trees. Oaks and hickories are both shade-tolerant species that produce relatively few seeds, yet their seedlings are able to remain in a stunted condition, sometimes for years, until a windfall or dying tree opens a gap in the forest canopy for them to fill. Unfortunately, the same characteristic makes oak and hickory seedlings particularly vulnerable to grazing by deer.

The woods has now entered the final stage of succession known as *climax vegetation*, a botanical term describing a diversified plant population that, in theory at least, will maintain itself indefinitely. In reality, however, some disturbance is always occurring. Trees are felled, if not by man, then by lightning, fire, or disease. As sunlight pours into the resulting clearing, poke and blackberries will appear as if by magic. Both have seeds that are able to hide in the soil for years, waiting for just such an opportunity.

The successional stages I have described are typical of an old field sequence in southeastern Pennsylvania. In other areas and under different circumstances the species may differ. For instance, in the South where the soil is dry and sandy, Virginia pine, *Pinus virginiana*, and pitch pine, *P. rigida*, are woody pioneers, often followed by oaks such as blackjack oak, *Quercus marilandica*, and chestnut oak, *Q. prinus*. Dry soils slow plant growth wherever they occur, and the rate of succession in such places is often slowed as well. Replacement occurs much faster on flood plains and

BOX ELDER
Acer negundo

bottomlands, where the soil is always moist. Here the early succession may be dominated by black willow, *Salix nigra*, box elder, *Acer negundo*, silver maple, *A. saccharinum*, and the climax by red maple, sycamore, *Platanus occidentalis*, swamp white oak, *Q. bicolor*, and black walnut, *Juglans nigra*.

The pioneer hardwoods of interior New England are usually birch, *Betula* spp., and white pine, *Pinus strobus*, followed by oaks and sugar maples, *A. saccharum*, with an eventual climax of hemlock, *Tsuga canadensis*, and beech, although much depends on the slope and exposure of the woods.

Particular species may vary, but a good rule of thumb is that the early successional stages everywhere are made up of sun-loving plants, a large proportion of which are aliens, with seeds that will not germinate in the shade. This is also true of early trees: young white pines cannot grow in the shade of old ones, for instance. Later successional stages are shade-loving, and the majority are also indigenous to eastern North America.

Between fifty and one hundred years after the field was abandoned, the climax vegetation will be in place. Nevertheless, a second century will have to pass before it will once again resemble the optimum North American deciduous forest. Throughout history and on every continent, human beings have destroyed the woods to make room for crops and herds. In many places they have destroyed the land as well, exhausting the fertility of the soil to the point of creating deserts where none existed before. Now that we are nearing the end of the twentieth century, expanding populations worldwide are putting intense pressures on ecosystems everywhere, but in spite of past abuse, our familiar eastern woods are still viable enough to contain a variety and abundance of plants unsurpassed anywhere. One hopes that we will have the wisdom and maturity to allow at least some of it to return, undisturbed.

· F · O · U · R ·

Plant Provinces

Of course, succession does not always end in forest. Bands of differing vegetation swirl across the North American map like marbled endpapers in an old first edition. Circling the globe just south of the polar ice cap is an expanse of rust and green land known as the Arctic Tundra. Beneath it is a band of deep forest green that sweeps in a giant parabola from southern Alaska across Canada, dipping over the U.S. border into the northern edge of Michigan, New York, and New England. This is the taiga, or Northern Coniferous Forest. It is mottled with sky blue lakes and moss green sphagnum bogs, called muskegs by the native tribes. Isolated drips of this forest green flow down the spine of the Appalachians as far south as western North Carolina and eastern Tennessee. A second coniferous band, called the Western Evergreen Forest, streams like a moire ribbon of differing tree associations from coastal Alaska into Mexico.

PRICKLY PEAR CACTUS
Opuntia humifusa

The eastern half of the United States is overlaid by a rich mosaic pattern known as the Eastern Deciduous Forest. It covers most of New England and the Piedmont country of the Atlantic states and spreads across the Appalachians to the Ohio River valley. It is the largest province within our borders as well as the most diverse, and its color varies with the seasons. Yellow-green in spring, it darkens to an older green in summer, exploding into the colors of fire every autumn before settling into an iridescent gray for winter.

South and east of the deciduous forest is the Coastal Plain. This is tidewater country, characterized by salt marshes, slow-moving rivers, pine forests, and cypress swamps. Not long ago, geologically speaking, it was the bottom of a shallow sea. The province includes much of what we still think of as the plantation country of the Old South, but it also extends around the base of the Appalachians into eastern Texas and up to the Mississippi River valley into southern Illinois.

Coating the tip of Florida is a piece of true Tropical Forest that also includes the Caribbean islands, coastal Mexico, and Central America. Members of the palm family, *Palmae*, are typical of this province. Seven genera are native to Florida, although they do not include the coconut palm, *Cocos nucifera*, or the date palm, *Phoenix dactylifera*, for both are introduced species. The ubiquitous red mangrove, *Rhizophora mangle*, an aggressive colonizer of Florida's estuaries and lagoons, is typical of tropical coastal vegetation. It was widely scorned as a pest tree until recent studies identified it as an important link in the complex food chain of the south Florida coast.

West of the Appalachian mountains, the eastern forest reaches in long, fingerlike peninsulas into the green and ocher stain that is called the Grasslands Province. These great prairies of the central plains are generally divided into the tall grass prairie of the Midwest, where rain is comparatively plentiful, and the drier short grass prairies of the Far West.

West of the prairies is the "cold desert" of the Great Basin Province of Utah and Nevada, and to the southwest the "hot desert" of the Sonoran Province of New Mexico and Arizona. Along the southern coast of California is the Chaparral Province or "hard-leaved forest," a type of dense

∴ PLANT PROVINCES ∴

scrub which develops in regions of mild, wet winters and very dry summers.

Of course, my map is obsolete. Two centuries of intensive civilization have reduced the remaining bits of native vegetation to isolated fragments. While a detailed examination of every community in every province is beyond the scope of one book, it is important to realize that they are plant communities on a grand scale, and that many of the same factors that determine their boundaries also determine the boundaries of smaller communities within them.

What are these factors? The simple answer is water, climate, soil conditions, and fire. Water, however, is not simply the amount of annual rainfall, for high temperatures also mean high evaporation rates, whereas at cool temperatures rainwater and snow are slow to disappear. Topography, too, influences drainage and the amount of water available to plants. It also makes a difference when the rain falls. Is it evenly divided throughout the year, or does most fall in one season? The term *climate* encompasses the annual rainfall and the range of seasonal temperatures plus the number of frost-free days (the length of the growing season). The varying hours of daylight and darkness are also important, for their shifting relationship is the signal that stimulates many plants to grow and bloom.

Soils everywhere are profoundly influenced by the underlying rock, but the size of the soil particles and the presence or absence of organic matter are also important. All of these affect the drainage as well as the pH factor, the relative acidity or alkalinity of the soil. It is the pH factor that affects the action of the soil bacteria that change nutrients into a form that is usable by plants. Fire, as we have seen, is the wild card. Its subtle effect on plant communities has been appreciated for relatively few years.

Consider the Arctic Tundra, where conditions are extreme. The annual rainfall is light, only about twelve inches a year, including snow. Yet only the top few inches of soil ever thaw. The subsoil is locked in permafrost, impeding drainage, and the low temperatures mean the rate of evaporation is slow. Consequently, the soil is both cold and saturated. Light is almost continuous during the brief arctic summer, but in winter the sun barely clears the horizon and the temperature averages zero or below. No trees grow on the tundra. Not only are their roots blocked by the permafrost, but trees must have at least eight weeks of temperatures above fifty degrees Fahrenheit in order to bloom and set seed. Tundra vegetation, therefore, is dominated by moisture-loving mosses, grasses, and sedges, although there are low-growing heaths, dwarf willows, and some herbaceous plants as well. Many of these burst into glorious bloom during the brief arctic summer.

It is an intriguing fact that in the temperate zone one can travel to northern climates by simply climbing mountains, and bits of Alpine Tundra can be found on the highest peaks of northern New England. Alpine Tundra, however, differs from true Arctic Tundra in a number of ways. In the mountains the summer days are shorter and the growing season longer. Snowfall is heavy and slow to melt. It is this, not permafrost, that inhibits the growth of trees, because the ground is never free of snow long enough for seeds to germinate and seedlings to develop. Moreover, drainage is usually good in the mountains, resulting in fewer mosses and more flowering plants.

Tundra rainfall, scarce as it is, stays. In the desert, what rainfall there is quickly drains away through the porous sandy soil and evaporates in the hot summer sun. Therefore, desiccation, or the excessive drying of their tissues, is the chief enemy of desert plants. There are also wide diurnal swings of temperature; it may climb well above a hundred degrees on a summer day, but drop as much as forty degrees at night. In the so-called cold desert, where winter frosts are not uncommon, sagebrush, *Artemisia tridentata*, is the dominant plant. Hot desert plants

∴ PLANT PROVINCES ∴

have fat, succulent stems and spiny leaves. They are often coated with a waxy surface to further protect them from desiccation. In the hot deserts of the Sonoran Province, cacti are abundant, as are agaves and yuccas. The desert is also famous for its myriads of bright annuals. During the brief periods of rain, they seem to spring from the earth to grow, bloom, set seed, and die within a few short weeks. While annuals are common in hot, dry climates, they decrease in abundance as one travels north, where the cold soils are not conducive to quick seed germination.

As we will see, our eastern sand dunes have many desert characteristics. Rain may be plentiful along the dunes and barrier beaches, but the sand is so porous that the water is only briefly available to plants, while the hot sun and ocean wind continually dry their tissues.

Grasslands tend to develop in the interior of large land masses, where the temperature variation is extreme and prolonged droughts are frequent. Grasslands, like deserts, are often in the "rain shadow" of high mountain ranges. As moisture-laden clouds sweep in from the ocean and are forced up over the heights, the cold air at the summit causes the water to condense and fall as rain or snow, leaving little for the land on the other side.

Prairie winters are cold and windy, but the buds of grass lie safely dormant at or below ground level, well mulched by the thatch of previous years. They spurt upward in the warm spring rains, but quickly revert to dormancy during the punishing droughts of summer. Prairie soil is more alkaline than that of the eastern woods, especially in the high plains of the western short grass prairie.

Grass evolved relatively late, probably in Eocene times, about sixty million years ago. Its appearance made possible the vast herds of bison and other herbivores as well as the busy populations of prairie dogs that once dotted the plains. Prairie dogs and bison, in turn, helped keep the grassland open. So did the native populations, who regularly set it afire to herd bison into circles for slaughter. Their fires, as well as those ignited by summer lightning, blew eastward on the prevailing winds, killing the woods and spreading the prairie into western New York and Pennsylvania. Now that both herds and tribes have disappeared, and the dense prairie sod has been cut by the plow, the woods have spread west once again.

The dense coniferous forest crosses southern Canada and dips into our northernmost states, clothing a land of low relief. Northern soils are sandy and acid under a deep layer of slowly decaying litter. They are called *podzols*, a Russian word for ash-gray soil, and typically contain a layer of white sand near the surface, left behind as humic acids percolate slowly downward. Podzol soils are not suitable for agriculture. That fact, plus the short northern growing season, has made clear-cutting by loggers the principal threat to northern forests. However, they are also threatened by acid rain, which substantially lowers the pH of the soil, interfering with the nutrients available to the trees. Thus weakened, they become vulnerable to insect damage and disease.

STARFLOWER
Trientalis borealis

Northern winters are long and cold, but again the rainfall is relatively sparse, about fifteen inches a year. The small, hard needles of coniferous trees resist desiccation, enabling them to survive where the evaporation rate is low. Farther south, grassland develops under the same conditions. Familiar spring wildflowers of the north woods are the goldthread, *Coptis groenlandica*, starflower, *Trientalis borealis*, and wood sorrel, *Oxalis montana*. The predominant tree species is the white spruce, *Picea glauca*. It is joined in the West by lodgepole pine, *Pinus contorta*, and in the East by balsam fir, *Abies balsamea*. Paper birch, *Betula papyrifera*, is everywhere, its gleaming white trunks are the grace notes that enliven the somber music of the evergreens. Tamarack, *Larix laricina*, and black spruce, *Picea mariana*, circle the frequent sphagnum bogs, while jack pine, *Pinus banksiana*, and aspen, *Populus tremuloides*, are the first to appear in burned-over areas.

As mentioned, bits of this boreal forest extend far to the south along the spine of the Appalachians, usually above five thousand feet. These patches are remnants of the time when the northern forest retreated southward before the ice sheets, and coniferous trees covered the entire landscape of the East. As the climate warmed, deciduous forest gradually replaced the evergreens at low altitudes, isolating the conifers high in the mountains. Over the ages, new species evolved so that now the dominant southern conifers are the red spruce, *Picea rubens*, and the fraser fir, *Abies fraseri*. These trees grow both higher and denser than their northern cousins because rainfall is plentiful in the southern mountains—up to one hundred inches a year in the Smokies.

All of the land now covered by the northern forest was buried repeatedly by glaciers, vast continental ice sheets that crept southward four times during the last million years. In North America, they spread from two centers: the Laurentic sheet in the East and the Cordilleran sheet in the West. The last advance was called the Wisconsin glacier. It covered the Great Lakes and all the adjacent land eastward to New York City, retreating about twelve thousand years ago.

A glacier forms when the difference between winter and summer temperatures is too small for the snow to completely melt. Year after year, the snowfall builds, eventually compressing into an enormous mound of ice. When the weight of the ice in the center of the glacier builds up enough to push out the edges, the glacier begins to move. The advancing ice penetrates cracks in the surrounding rock, breaking off large chunks that may then be transported great distances on the moving ice.

The Wisconsin glacier advanced across the landscape like a monumental plow, killing all the vegetation and moving vast piles of earth before it. These materials, called moraines, were often left behind when the ice finally melted, forming islands and land masses where none had existed before. The glacier diverted the course of rivers, carved mountains, mounded hills, and altered the level of the sea. The crushing weight of the ice even compressed the land surface, which then sprang back unevenly, leaving poorly drained depressions. Sometimes isolated blocks of slow-melting ice remained behind while deposits left by the retreating glacier piled up around them. When these blocks melted, they formed "kettle holes" which, when filled with water, formed the deep, clear lakes of the north.

The glacier forced the vegetation to move south ahead of it. Because the ancient Appalachian ridge runs north and south, the plants had plenty of room to retreat and were thus saved from the fate of numbers of European flora that, too tender to migrate over the east-west ridges of the Alps, were lost forever. This is one reason that the native European flora is far less diverse than ours. Centuries of human occupation is another.

As the climate warmed, the deciduous trees began to migrate north once again, a process that is still continuing. Deciduous trees cannot survive in the fierce, dry cold of the northern woods. They need about forty inches of rainfall annually, spread evenly throughout the year. Even so, they must drop their leaves in winter when the groundwater is frozen and protect their incipient buds with waterproof scales.

Today the Eastern Deciduous Forest covers most of the Appalachians and their foothills from New England to Georgia. It is hard to imagine just how old the Appalachian land mass really is. Geologists say as much as five hundred million years. If so, it was there before life crawled out of the sea, a sea that then washed over the central part of this continent. Three hundred and fifty million years ago, the New England mountains pushed up at the northern edge of

∴ PLANT PROVINCES ∴

Appalachia and the first lichens and algae covered the earth.

During the next one hundred million years, the seas withdrew, leaving vast swamps in which enormous plants called *Arthrophyta* flourished. The field horsetail, *Equisetum arvense*, a common plant of railroad embankments and wet, gravelly soils, is one of its descendants. *Lepidophyta*, or scale trees, were also prevalent in prehistoric times. They have evolved downward, so to speak, into our running ground pines, *Lycopodium* spp. All these plants were prevalent when the dinosaurs roamed the earth. By the end of their time, one hundred and thirty million years ago, evolution took one of its startling leaps forward. Insects appeared, then flowering plants, and finally mammals. All three were to inherit the earth.

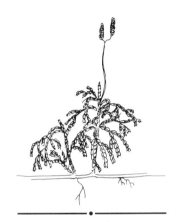

RUNNING PINE
Lycopodium complanatum

At that time the climate was apparently milder than it is today, and tropical forests advanced as far north as the state of Washington. During the next fifty million years, the western mountains, which had first appeared in Jurassic times, one hundred and fifty-five million years ago, rose to new heights and the climate cooled at high altitudes, causing a rain shadow on the eastern side of the range that forced the tropical forest to retreat southward. During the Miocene period, twenty-eight million years ago, a temperate forest moved southward from Alaska and northern Canada, stretching down across the continent from Oregon to Appalachia. Temperate forests also covered Europe and Asia, a worldwide band of trees that botanists have named the Arcto-Tertiary Forest. In Europe, as we have seen, it was largely killed off during the Pleistocene Ice Age, but in Asia, where the mountain ranges also run north and south, more plants were able to survive the onslaught of the glaciers. Today the numerous genera that are common to both continents are survivors of that vast Arcto-Tertiary Forest.

In Asia, the woods have been cut repeatedly over the centuries and very little remains. In the cove forests of the

Smokies, however, one can still walk among trees so venerable they easily recall that ancient time, and everywhere our eastern forest is filled with the young descendants of Arcto-Tertiary trees. Of course, many species have been lost and many others added over millions of years, yet today, even after two centuries of human punishment, our woods are still a rich mosaic of plant associations, each shaped by exposure, soil, and drainage. In the next chapter we will examine some of them more closely.

II

Plant Communities

· F · I · V · E ·

Deciduous Woods

TULIP POPLAR
Liriodendron tulipifera

This discussion of deciduous woods is, obviously, vastly oversimplified. Our eastern woods are an extremely complex plant community, consisting of hundreds of different species of trees, shrubs, and herbaceous plants in many different associations. The problem is further complicated by the fact that most woods are still in some phase of woody succession. Even so, they still contain a bigger variety of species of trees than are found anywhere else in the northern hemisphere.

THE MIXED MESOPHYTIC FOREST

The tulip poplar, *Liriodendron tulipifera*, one of the first species to dominate the forest in many parts of the East, is a tree of ancient ancestry and erect bearing. It is a member of the magnolia family. Flowers of the magnolia family are large and waxy with numerous styles surrounded by numerous stamens. Because they are so unspecialized, they are considered by some botanists to be early on the evolutionary scale. This is not to say, at least in the case of the tulip tree, that they are inefficient. Tulip trees are spry opportunists. They spring up in every available clearing and

SEED POD OF TULIP POPLAR
Liriodendron tulipifera

quickly rise to the tallest level. They must to survive, because their seeds do not sprout in the dense shade of a mature forest. Young trees bloom when they are about fifteen or twenty years old, holding their flowers high in the sunshine, balanced like green and orange candle holders at the end of the branches. The seed pods look like wooden many-petaled flowers and stay on the branches most of the winter, making the tree easy to identify from afar.

Tulip poplars are trees of deep, moist soils. They are conspicuous members of the mixed mesophytic forest, a botanical mouthful meaning a forest with no single dominant species growing in rich, well-drained soil. The richest mixed mesophytic forest grows in the southern Appalachians and is the evolutionary parent of all the other deciduous tree associations found in the eastern woods. It is the modern-day descendant of the great worldwide temperate forest of Tertiary times. Trees endemic to the southern Appalachians that enliven the woods there, as nowhere else, include the silverbell, *Halesia monticola*. When it blooms in April, it seems to sprinkle the treetops with white confetti, so numerous are its bell-like flowers. The Carolina silverbell, *H. carolina*, is a smaller species that is eagerly sought after by gardeners. Both are hardy as far as eastern Massachusetts. The sweet buckeye, *Aesculus octandra*, is also found only in these southern mountains. It is part of several different forest communities at various elevations and resembles the alien horse chestnut, *A. hippocastanum*, often planted in the Northeast. The sweet buckeye has the typical palmately compound leaves of the genus, the leaflets surrounding the center like spokes of a wheel. The greenish yellow flowers are held on showy, upright spikes during May and June.

Two deciduous magnolias also grow in these southern forests: the umbrella magnolia, *Magnolia tripetala*, a tree with giant, almost tropical-looking leaves, and the Fraser magnolia, *M. fraseri*, distinctive for its "ears" at the base of

CAROLINA SILVERBELL
Halesia carolina

∴ DECIDUOUS WOODS ∴

each leaf. Many southern plants have *fraseri* for a specific name. They are named for John Fraser, a Scottish collector of American plants who lived from 1750 to 1811 and kept a nursery in Chelsea, England. He made several expeditions to North America, collecting plants from Newfoundland to the Carolinas.

Other great trees of this forest include the sugar maple, *Acer saccharum*, the beech, *Fagus grandifolia*, and the hemlock, *Tsuga canadensis*, all familiar trees in New England. Also present are white ash, *Fraxinus americana*, and the black, white, and red oaks, *Quercus velutina*, *Q. alba*, and *Q. rubra*, which grow in mesic soils all over the East. The red maple, *Acer rubrum*, is here, too, with its deep carmine flowers in earliest spring and brilliant red foliage in autumn. It is a tree found in nearly all our woods, apparently able to adapt to a variety of habitats. Here too is the black cherry, *Prunus serotina*, known to most of us as an opportunistic tree of thickets and hedgerows.

Along stream banks and bottomlands in the mixed mesophytic forest grow many trees that are also found in the low forests of the Coastal Plain and along riverbanks in the Piedmont: the black walnut, *Juglans nigra*, supplier of sweet nuts and cabinet wood, and the mottled sycamore, *Platanus occidentalis*, as well as black willow, *Salix nigra*, and sweet gum, *Liquidambar styraciflua*. This last is a southern tree with star-shaped leaves as clean and crisp as a Christmas cookie, and dangling seed pods as prickly as a burr.

These are trees of the canopy, a botanist's name for the tallest trees in a forest community. Typically, a woods has four levels. Below the canopy are the understory trees, smaller species that tolerate the shade of the arboreal giants. Then come the shrubs, and finally a groundcover of herbaceous plants and ferns. Lowest of all are the mosses. Each layer has adapted to progressively diminishing levels of light, so herbaceous plants, which need the least, are both

SWEET BUCKEYE
Aesculus octandra

FRASER MAGNOLIA
Magnolia fraseri

DOGWOOD
Cornus florida

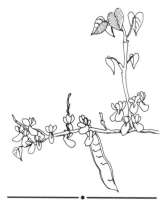

REDBUD
Cercis canadensis

the last to arrive in a developing woods and the first to green up in the spring.

The tallest tree of the understory is the black gum or tupelo, *Nyssa sylvatica*, although some individuals may grow large enough to reach canopy level. Its lustrous leaves catch and reflect the filtered sunlight and turn blood red in early autumn—the first in the forest to change. Tupelo is found throughout the East on dry, sandy soils as well as rich bottomlands. Botanists recognize a second variety, *N. sylvatica biflora*, which grows in heavy, wet soils on the Coastal Plain.

Shadbush, *Amelanchier* spp., unfolds its lacy blooms in earliest spring, blooming along the banks of streams, often growing with two members of the birch family, Betulaceae, the American hornbeam, *Carpinus caroliniana*, and hop hornbeam, *Ostrya virginiana*. American hornbeam has bark like twisted licorice, while the bark of hop hornbeam peels off in cinnamon brown scales. In May the most conspicuous tree of the understory is the flowering dogwood, *Cornus florida*, an especially lovely sight when mixed, as it often is south of Maryland, with redbud, *Cercis canadensis*, in drifts of pink and white.

The shrub layer of the mixed mesophytic woods is a subtle parade of bloom beginning in early April when the spicebush, *Lindera benzoin*, flowers like a faint wash of yellow among trees still bare of foliage. By May, the mapleleaf viburnum, *Viburnum acerifolium*, blooms with pinxterbloom, *Rhododendron periclymenoides*, the exquisite wild azalea of the eastern woods. In June the white flowers of elderberry, *Sambucus canadensis*, billow like foam along stream banks, and in August the enticing fragrance of the sweet pepperbush, *Clethra acuminata*, perfumes the woods. The final curtain falls in mid-November when the improbable straplike flowers of witch hazel, *Hamamelis virginiana*, gleam among the falling leaves.

Vines are common in this forest as well, notably the

∴ DECIDUOUS WOODS ∴

Virginia creeper, *Parthenocissus quinquefolia*, and various species of wild grape, *Vitis* spp. Spiny greenbrier, *Smilax* spp., grows thick at the forest edge, and the great hairy ropes of poison ivy, *Rhus radicans*, climb the ancient trunks. All parts of the poison ivy plant contain an oil that will cause an irritating rash in susceptible people. The plant takes many forms. It may be a high-climbing vine, a groundcover, or a thick, shrubby thicket two to three feet high. It does not twine like honeysuckle or bittersweet but clings to tree bark by means of holdfasts. In autumn, the shiny, three-part leaves turn a brilliant red, and clumps of dry, white berries cling to the plant well into winter, providing valuable food for birds.

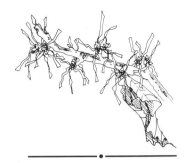

WITCH HAZEL
Hamamelis virginiana

Ferns are everywhere. The evergreen Christmas fern, *Polystichum acrostichoides*, is among the first to unroll its crosier in spring. Cinnamon fern, *Osmunda cinnamomea*, is easily identified by its fertile "cinnamon stick" that rises in the center, surrounded by sterile leaves, and New York fern, *Thelypteris noveboracensis*, has leaves that taper to a narrow point at both top and bottom. Interrupted fern, *Osmunda claytoniana*, may grow five feet high in swampy areas. Its leaves are conspicuously "interrupted" by brown fertile pinnae in the middle of the stem. These are but a few of the more easily identified species. There are many more.

HEPATICA
Hepatica americana

Woodland flowers begin with skunk cabbage, *Symplocarpus foetidus*. Its mottled purple hood seems to appear overnight on moist bottomlands during the first warm days of March. Hepatica, *H. americana* or *H. acutiloba*, unfolds its delicate lavender blossoms soon after, often, like many evergreen species, on cool northern slopes where winter fluctuations of temperature are minimal. A rare flower of early spring in the Carolina mountains is oconee bells, *Shortia galacifolia*, which blooms among the laurel and rhododendron. Like Franklinia, *shortia* was once one of the lost plants of America. It was originally found in 1787 by the French botanist André Michaux, one of the most adventur-

∴ FIELD AND FOREST ∴

OCONEE BELLS
Shortia galacifolia

TROUT LILY
Erythronium americanum

ous travelers and highly trained scientists of the eighteenth century. Michaux didn't name his find, but took a herbarium specimen home with him to Paris. Michaux died in 1803, and his collection languished unattended until the American botanist Asa Gray went to France in the 1830s. Intrigued by the little plant, he named it *Shortia galacifolia* after Charles Wilkins Short, a fellow botanist from Kentucky. At that time Gray knew only that it grew somewhere in North America. It wasn't until a similar Japanese species, *S. uniflora*, was found and described that he had a clue to its possible habitat. Even so, *shortia* was not found again in the wild until the 1880s, when Gray was an old man in the last years of his life. It is still one of our rarest wild plants.

Many fleeting flowers of the forest floor are called "spring ephemerals" because by early summer they have not only finished blooming, but already set seed and relapsed into dormancy. Plants need light in order to flower, and these species have evolved so as to take advantage of the warm spring sunlight at the only time it is available to them, before the leaves of forest trees have developed. Trout lilies, *Erythronium americanum*, whose mottled leaves and nodding yellow flowers appear along stream banks in April, are spring ephemerals. So is the fragile spring beauty, *Claytonia virginica*, as well as Dutchman's breeches, *Dicentra cucullaria*, toothwort, *Dentaria* spp., and anemone, *Anemone quinquefolia*. All of these carpet the slopes of a mature forest. Scattered among them is the delicate rue anemone, *Anemonella thalictroides*, and the waxy three-petaled flowers of *Trillium* spp. Later, May apples, *Podophyllum peltatum*, jack-in-the-pulpit, *Arisaema atrorubens*, Solomon's seal, *Polygonatum biflorum*, and plume flower, *Smilacina racemosa*, appear. These are not spring ephemerals, for the plants remain all summer, setting fruit by late summer or fall. (Incidentally, jack-in-the-pulpit is only a "jack" while the plant is young. The tiny flowers that are contained within the hood, or spathe, start out as male and change to female

∴ DECIDUOUS WOODS ∴

as the plant matures. The plants that ultimately set fruit are "jills-in-the-pulpit"!)

These flowers fade as the tulip trees and magnolias bloom above them. By mid-June the spring ephemerals are gone and most tree growth is finished as well. The rest of the season is spent developing seeds and buds for the following year.

JACK-IN-THE-PULPIT
Arisaema atrorubens

OAK-CHESTNUT FOREST

Among the evolutionary offspring of the mixed mesophytic forest is the oak-chestnut forest. This is the principal mesic forest of the Piedmont, that area of rolling countryside east of the Appalachians and west of the fall line (the term *fall line* originally referred to the waterfalls that marked the upper limit of navigation on the major rivers of the East). The name oak-chestnut is an obvious misnomer now, because the American chestnut, *Castanea dentata*, has disappeared from our woods. Virtually all the trees were killed by a fungus bark disease believed to have been introduced from Asia about 1900. From time to time sprouts still appear from remaining stumps, but the young trees rarely reach flowering size before succumbing to the blight. Botanists hope that a strain of the same fungus found in Europe that infects and inactivates the pathogenic strain will eventually make it possible to have chestnut forests once again. So far the method has been used successfully on individual trees, but the new strain has failed to spread widely on its own. So, for our time at least, the American chestnut is gone. Hickories have increased to fill the gap, and some botanists now call this the eastern oak-hickory forest, to differentiate it from the oak-hickory forest that grows in dry soil on the western slopes of the Appalachians.

The difference between the eastern oak-hickory forest

and the mixed mesophytic forest is principally in the number of species found in the canopy. As one travels north, many of the trees endemic to the southern mountains, such as the magnolias and silverbells, drop out of the woods' population. The tulip poplar is still abundant, however, and is quick to appear in a developing woodland or in a space left by a fallen tree. It quickly ascends to canopy height and persists in the mature forest as a large tree, often surrounded by smaller, later arriving oaks and hickories. White ash is also common in these woods, and red maple is everywhere, often growing in combination with beech on north-facing slopes, but the sugar maple is now found only on floodplains and bottomlands. Most of the understory trees, shrubs, and wildflowers of the mixed mesophytic woods are also abundant in this forest.

The oaks, central to many woodland communities, are a large and unruly clan. There are fifty-eight species of oak in North America, excluding Mexico, which has one hundred and fifty species of its own. In fact, our American oaks are thought to have migrated from Mexico thousands of years ago, spreading northward as the climate warmed. Southern oaks tend to be evergreen, a carryover from their tropical heritage, and even deciduous northern oaks hang on to their leaves well into winter.

Oak trees have deep penetrating roots, and the genus has adapted to a wide range of soil conditions. Chestnut oak and blackjack oak, *Quercus prinus* and *Q. marilandica*, are trees that grow on the driest soils of our region. Black oak, *Q. velutina*, likes moderately dry conditions, while white oak, *Q. alba*, and red oak, *Q. rubra*, are trees of the mesic forest. Willow oak, *Q. phellos*, water oak, *Q. nigra*, swamp white oak, *Q. bicolor*, and swamp chestnut oak, *Q. michauxii*, all grow in wooded swamps and wet bottomlands.

Generally, however, oaks are more tolerant of dry conditions than many trees, and there is evidence that where the land has been badly eroded or forests excessively cut,

DECIDUOUS WOODS

or where water tables have been depleted by overdevelopment, oaks tend to supplant other species, sometimes forming nearly pure stands. The area covered by oak forest has indeed increased since the early years of this century, but now the oaks in turn are being threatened by the rapid spread of the gypsy moth and exploding populations of white-tailed deer that feed on the seedlings.

For a species that calls up the image of endurance, durability, and strength, oaks are notoriously slippery to identify. Botanical descriptions of oak trees are full of phrases such as "somewhat hairy," "somewhat thickened," or "usually but not always." To further confuse the amateur, leaves on young oak trees are very variable and may not resemble those on mature specimens of the same species.

Oaks are members of the beech family, Fagaceae, an important group of timber trees for the wood is hard and durable and takes a good finish. Taxonomists have divided the oaks into two groups, the white and the red (also called the black). The white oak group have rounded lobes without bristle tips; their acorns mature in one year and are found on the current season's growth. The inner surface of the acorn shell (not the cup) is smooth, not hairy. Two familiar members of the white oak group are the white oak, *Q. alba*, a dominant member of the eastern oak-hickory forest, and the chestnut oak, *Q. prinus*, a tree of especially dry soils.

The trees of the red (or black) oak group have bristle tips on their leaves and acorns that take two years to mature so that both tiny new ones and larger old ones are frequently found on the tree in summer. The inner surface of the acorn shell is hairy. Members of the red oak group commonly found in the oak-hickory forest are the red oak, *Q. rubra*, and the black oak, *Q. velutina*, both large stately trees, and the scarlet oak, *Q. coccinea*, which is usually smaller. All three species apparently hybridize, producing

WHITE OAK
Quercus alba

many individuals that vary from the textbook descriptions. Common farther south is the Spanish oak, *Q. falcata*, also called the southern red oak. Mature Spanish oaks are large, handsome trees with distinctive leaves that are glossy on top but covered with orange fuzz beneath, making them conspicuous from afar, especially on a windy day.

Hickories are members of the walnut family. They have deep taproots that enable them to adapt well to dry soils. Hickories usually appear in a developing woods after the oaks have begun to bear acorns, because the acorns attract squirrels which then bring in the hickory nuts. Prevalent among the hickories found in this community are the mockernut, *Carya tomentosa*, its common name apparently referring to the impossibly thick shell that surrounds a tiny amount of meat within, and the all but inedible bitternut, *C. cordiformis*. Less common is the shagbark hickory, *C. ovata*, whose long strips of peeling bark make it easy to identify. Shagbark has the sweetest nuts of all. They were once a staple food of the native populations, who ground and mashed them, mixing the paste with water to make a "hickory milk" that was used in cakes. A fourth widespread hickory is the pignut, *C. glabra*, the nuts of which were once commonly fed to swine.

Where the soil is particularly dry, the character of the forest changes. This is often apparent on the tops of ridges and hills where the rock substrata is close to the surface and the topsoil thin, but it is also characteristic of soils that have been badly eroded from poor management in the past. Individual trees become smaller in the dry woods for the lack of moisture inhibits their growth, and there are many fewer species in the community.

Certain species of oak are always indicative of dry soils: the post oak, *Q. stellata*, the chestnut oak, the bear oak, *Q. ilicifolia*, and the more southern blackjack oak, *Q. marilandica*. They are sometimes mixed with dogwood and the adaptable red maple. New Jersey tea, *Ceanothus americanus*,

MOCKERNUT HICKORY
Carya tomentosa

∴ DECIDUOUS WOODS ∴

is a conspicuous shrub of dry woods. So is *Rhus aromatica*, the fragrant sumac, a plant very different in appearance from the familiar sumacs of open fields. Particularly diagnostic of dry woods, however, is the presence of ericaceous shrubs, or plants belonging to the heath family, Ericaceae. One of these is sourwood, *Oxydendrum arboreum*, a particularly attractive small tree that is found on ridges south of Pennsylvania. Sourwood has white flowers very like those of the Japanese andromeda, *Pieris japonica*. They form feathery sprays in July, conspicuous against the shiny foliage, and as an extra dividend, the seed pods continue to cling to the tree while the leaves turn a brilliant coppery red in October. All these features make sourwood a favorite small tree for the garden.

SOURWOOD
Oxydendrum arboreum

Mountain laurel, *Kalmia latifolia*, is probably the most well-known member of the heath family. It is common along the ridges of the Shenandoah where it nestles among stunted oaks with trunks green with lichens and branches frequently broken by ice and wind. Mountain laurel has clusters of intricate pink and white blossoms, each flower like a tiny hoop petticoat. In each the stamens are tucked into pockets in the corolla, ready to be popped free by a pollinating bee. Mountain laurel also grows freely around stony ledges in New England as well as in the low oak and pine forest of the Coastal Plain.

MOUNTAIN LAUREL
Kalmia latifolia

Various species of blueberries and huckleberries, genera *Vaccinium* and *Gaylussacia*, are also members of the heath family that are plentiful in dry oak forests. They have flowers like dangling pink or white bells as well as edible fruit. Wintergreen, *Gaultheria procumbens*, is a low creeping heath that is also found in these woods. It has waxy, bell-like flowers in summer and leaves that yield the fragrant oil of wintergreen. The pealike red berry is fragrant, too, and edible besides.

WINTERGREEN
Gaultheria procumbens

Herbaceous plants of the dry oak woods might include snakeroot, *Eupatorium rugosum*, wild columbine, *Aquilegia canadensis*, and sarsaparilla, *Aralia hispida*. Sarsaparilla was another plant in vogue in the eighteenth century, because it was thought to be a cure for venereal disease. At various times it has also been made into a beer, smoked as a remedy for asthma, and used as a flavoring for soft drinks.

Oaks and hickories are prevalent in the woods of southern New England, especially near the coast, but the principal community of interior New England is the hemlock, white pine, and hardwood forest.

Hemlock, White Pine, and Hardwood Forest

BUNCHBERRY
Cornus canadensis

This is a forest of cool, moist climates. It is young in terms of evolution, because all of the area it now covers was once buried beneath the glaciers. Hemlock, *Tsuga canadensis*, and white pine, *Pinus strobus*, appear as patches of darkest green on the hillsides, mixed with hardwoods like sugar and red maple, yellow birch, *Betula lutea*, and white ash, *Fraxinus americana*. Red oaks and beech, *Fagus grandifolia*, are also common. The shrubs include many we have already mentioned, plus hobblebush, *Viburnum alnifolium*, yew, *Taxus canadensis*, and alternate leaved dogwood, *Cornus alternifolia*, a graceful cousin of the familiar flowering dogwood. Elderberry and witch hazel are also frequent. Ferns are plentiful and flowers include Canada mayflower, *Maianthemum canadense*, also called wild lily of the valley, besides many of the mesic varieties of the eastern oak-hickory woods. A particularly choice flower of northern New England is a tiny, ground-hugging dogwood called bunchberry, *Cornus canadensis*.

With the exception of coastal marshes and inland bogs, the New England forest was filled with ancient trees, many with trunks several feet in diameter, when the Pilgrims arrived in Plymouth in 1620. Wide boards taken from these trees may still be seen in early houses. Two hundred years later, the forest had been virtually eliminated. By 1850, only 5 percent of New England was still covered with trees, so the oldest woods that exist today are little more than a century old. Some botanists think that pines and hemlocks were more common in the early days than they are today, because, unlike hardwoods, they do not resprout when cut. All of today's conifers, therefore, had to compete as seedling trees.

THE COASTAL PLAIN FOREST

East of the Piedmont fall line is the Coastal Plain, a table-flat area of former sea bottom only a few feet above sea level. There is no Coastal Plain in New England; the rocky soil there was largely deposited by the glaciers and is quite different from the sandy sediment left by the sea. The boundary that marks the limit of the ice is evident as you travel south of New York City. The Coastal Plain spreads down the Atlantic coast from a narrow strip in northern New Jersey, continuing around the base of the ancient Appalachian continent into eastern Texas.

Because the Coastal Plain is prime agricultural land, few original woods remain. Those that do have been managed to such an extent that it is unclear what climax forest would develop, given the time and opportunity. Pines are common throughout. Pines, as gymnosperms, are of far more ancient origin than deciduous trees. Fossil records show that conifers much like those of today began appearing in the Triassic period, one hundred and eighty-five million years ago. Southern pines are quite unlike the glaucous, feathery white pine of the New England woods. They are stiff of habit and bristly of character. Some are quite contorted. The species most common on the Delmarva Peninsula, a representative area of Coastal Plain communities, are pitch pine, *Pinus rigida*, scrub pine, *P. virginiana*, and loblolly pine, *P. taeda*. Common farther south are the shortleaf pine, *P. echinata*, and the longleaf pine, *P. palustris*. The last is known colloquially as "broom pine" because its needles, which are frequently over a foot long, were once used to make brooms.

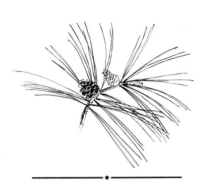

LOBLOLLY PINE
Pinus taeda

Pines thrive on sandy soils that are less suited to deciduous forest, and they are also found in conjunction with various species of the dry oak forest. We will examine this community more closely in the chapter on pine barrens.

The deciduous woods of the Coastal Plain contain many trees found along stream banks and bottomlands within the mixed mesophytic forest, such as sycamore, *Platanus occidentalis*, black willow, *Salix nigra*, sweet gum, *Liquidambar styraciflua*, and black gum, *Nyssa sylvatica*, here the variety *biflora*. These woods are also full of plants of proven garden merit. The American holly, *Ilex opaca*, is common among the pines, particularly near the coast. Both the sweet bay magnolia, *Magnolia virginiana*, and that stately tree of southern gardens, the bull bay, *M. grandiflora*, are trees of the Coastal Plain. The sweet-smelling swamp and coast azaleas, *Rhododendron viscosum* and *R. atlanticum*, also grow here. So does the winterberry, *Ilex verticillata*, noted for its brilliant berries at Christmastime. Other shrubs include the shrubby wax myrtle, *Myrica cerifera*, a southern evergreen cousin of the bayberry, *M. pensylvanica*. A distinctive small tree called Hercules' club, *Aralia spinosa*, a relative of sarsaparilla, frequently grows near the forest edge. It somewhat resembles a sumac, but has numerous prickly spines on its trunk and branches, and huge, creamy white clusters of flowers in August.

SWEET BAY MAGNOLIA
Magnolia virginiana

∴ DECIDUOUS WOODS ∴

We have already seen how higher altitudes favor northern or boreal plants. Similarly, the warm ocean current known as the Gulf Stream moderates the coastal climate so that many southern, or austral, plants reach the northernmost part of their range along the Coastal Plain. Loblolly pine is one of these. (*Loblolly* is a southern country word meaning mudhole or swamp.) Pond pine, *P. serotina*, is another southern species that reaches its northern limit in the New Jersey Pine Barrens. So does water oak, *Q. nigra*, and its frequent companion, sweet gum. Blackjack oak, *Q. marilandica*, and willow oak, *Q. phellos*, are also southern species that extend north along the Coastal Plain. Persimmon, *Diospyros virginiana*, is another. Persimmon, a close relative of ebony, is a tree of primitive origin that dates from the early Cretaceous period, one hundred and thirty million years ago.

Woodland flowers are not as plentiful on the Coastal Plain, perhaps because truly mature forests are rare. However, numerous species of ground pines and club mosses, *Lycopodium* spp., carpet the forest floor.

· S · I · X ·

Open Lands

Forests can provide fuel and shelter, but agriculture must take place in the sun. So for thousands of years the development of civilization has resulted in the cutting of the forest. Forests and agricultural land are valued in densely populated Europe, where little wildness remains, and have been meticulously managed for maximum fertility and economic production for centuries. Yet this was not always the case. In the ancient civilizations of the Mediterranean, the deforestation was so extensive and the land so badly abused that soils that were fertile in Classical times are now too eroded and poor to support the original vegetation. Unfortunately, Europeans coming to eastern North America found such a vast expanse of land that was both fertile and available that they quickly reverted to the earlier pattern of deforestation and abuse.

The woodland tribes of eastern North America were not an agricultural people. They lived primarily by hunting and fishing, although they did grow corn, beans, pumpkins, and squash in clearings near their villages. When the soil was exhausted, they simply abandoned the plot and cleared another, but because the population was small and the villages widely scattered, there was little effect on the woods as a whole.

The European settlers, on the other hand, immediately set to work clearing land, putting up fences, and establishing farms. Within two generations, they had changed the whole character of the American landscape. The first land to be cleared was always mesic land: rich, moist, and well drained, for the land that grew the richest forest also grew the best crops. Early accounts by Europeans visiting this country invariably remark on the poor farming practices of the American farmer, complaining that they neither rotated their crops nor manured their fields. We know that crop rotation was not common in Pennsylvania until the 1780s, although it had been practiced in England for over a century. But farm labor was both scarce and expensive in the New World, whereas land was there for the taking. There seemed to be no reason to husband it, so most American farmers simply cleared "fresh" land as it was needed. In Pennsylvania, the average eighteenth-century farm was about one hundred and twenty-five acres. As much as thirty acres of this might lie fallow at any one time, often in what was called *bush fallow*. A revealing term: it was probably exhausted land that had not been cultivated for several years.

Only about twenty acres of the typical Piedmont farm was left in woodland, usually on slopes too steep to plow. They were probably poor woods in any case, in terms of their plant life, for in those days it was common practice to let cattle and sheep graze in them. Woods became so scarce that as early as the 1790s Philadelphians were remarking on the lack of firewood, complaining that it had to be transported great distances to supply the city's needs.

About thirty acres of these early farms were kept as pasture or hayfields, while orchards and gardens were planted near the homesteads. The colonists brought over the seeds of forage grasses, fruit trees, vegetables, and medicinal herbs as well as many garden flowers. Many weeds came as well. Adapted over many thousands of years to disturbed soil conditions, they soon proliferated in the newly cleared land of the New World with little competition from a native flora adapted to a forest habitat.

Many of our alien field flowers of today date from those early times. Flax, for example, *Linum usitatissimum*, was an important field crop in early Pennsylvania, grown for its fiber and oil. It escaped to grow along modern roadsides. Queen Anne's lace, *Daucus carota*, is really a carrot. It and dandelion, *Taraxacum officinale*, were grown as vegetables in New England before 1700. Indeed, dandelions have been an important food plant for centuries, and directions for growing them were given in several early illustrated books called herbals. In 1871, four varieties of dandelion were exhibited at the Massachusetts Horticultural Society, and by 1884 a Mr. Corey of Brookline was growing them from seed for the Boston market. Nearly every part of the plant is edible, and the leaves have more vitamin A than most greens that are commonly cultivated today. However, they tend to be bitter unless picked very young, and must be cooked in two or three changes of water.

Tansy, *Tanacetum vulgare*, yarrow, *Achillea millefolium*, and feverfew, *Chrysanthemum parthenium*, were grown as medicinal herbs. Flannel mullein, *Verbascum thapsus*, was used for everything from curing warts to dyeing hair, while teasel, *Dipsacus sylvestris*, was grown for its prickly seed head, which was used to card wool. Butter and eggs, *Linaria vulgaris*, common buttercups, *Ranunculus acris*, and field daisies, *Chrysanthemum leucanthemum*, are all on early garden lists, apparently brought by homesick colonists to remind them of home. The knapweeds, *Centaurea* spp., were also brought as garden flowers. We have only one American species of knapweed, *C. americana*, and it is a native of the Southwest. Celandine, *Chelidonium majus*, and bouncing bet, *Saponaria officinalis*, are also on the early lists. The tawny day lily, *Hemerocallis fulva*, that brightens our roadsides in July, was nicknamed the "outhouse flower" by midwestern settlers because it grew so luxuriously in that location. How it has spread so far is a botanical mystery because it has been identified as a triploid form called

FLANNEL MULLEIN
Verbascum thapsus

"Europa" that never sets seed! Apparently the smallest root segment will suffice for a new plant to develop.

Most of the invasive prickly thistles of our meadows and fields are also aliens, possibly arriving as stowaways in shipments of grain. The Canada thistle, *Cirsium arvense*, which is not Canadian but European, is the most aggressive of the lot. It is a perennial that spreads by underground stolons as well as by far-flying thistledown, and can soon take over a field. It was such a pest of newly cleared land that Vermont passed the first antithistle law in 1795, requiring farmers to eliminate it. Bull thistle, *Cirsium vulgare*, and nodding thistle, *Carduus nutans*, can also become pests in fields and meadows. They are biennials, however, and do not have invasive root systems. A good way to control them is by mowing or by cutting off the flower with a weed whacker before it has a chance to go to seed.

We do have several native thistles, all of which are more choosy in their habitats. The three that might be found in fields or meadows belong to the genus *Cirsium*, which differs from the genus *Carduus* by the tiniest botanical technicality having to do with the hairs on the pappus, or modified calyx. Pasture thistle, *C. pumilum*, is our largest flowered species with heads two or three inches in diameter. Field thistle, *C. discolor*, is distinguished by a felt of white wool on the undersides of the leaves, while tall thistle, *C. altissimum*, lives up to its name by growing up to twelve feet tall.

FIELD THISTLE
Cirsium discolor

The new settlers also grew native American flowers, many of which later traveled back to England to be hybridized as garden flowers. Thus, our New England and New York asters became England's Michaelmas daisies. The colonists also grew sunflowers and wild geraniums as well as hepatica, bloodroot, lilies, lobelias, iris, and violets.

Plant communities of open fields and meadows are dominated by grasses, legumes, and flowers of the daisy family, Compositae, which are also called Asteraceae.

GRASS

It is odd that people pay so little attention to grass. Not only is the grass family, Gramineae, the largest family of flowering plants worldwide, but it is also the most useful to man. Grasses include all the grain crops: wheat, barley, rice, oats, rye, and Indian corn. Of these, only wild rice and Indian corn are native to the Americas. Grasses mixed with legumes make up the forage that feeds our livestock. Grass stems have been used to thatch roofs, weave baskets, and bed animals. Lacking trees, the early Nebraska pioneers even built houses from grass sod. In nature, grasses are important because they are the first plants to clothe barren land, preventing wind and water erosion.

Grasses are monocots, a subdivision of flowering plants. When the seed of a monocot sprouts, it has only one seed leaf. Monocots, which include, among others, the lily and iris families, also have straplike leaves and flower parts in multiples of three.

Grass flowers are wind-pollinated, and except for the heavier grains, their seeds are small and easily dispersed. The entire flowering part of a grass plant is called the *ear*, while each individual flower is call a *spikelet*. Grass stems, called *culms*, are usually hollow, but they have solid joints, called *nodes*. The leaves, which are attached at the nodes, consist of a sheath enclosing the culm and a blade that is long and narrow. Grasses have persistent, adaptable roots that spread out in damp weather and bore downward in dry. The roots of native prairie grasses can store enough food to outlast years of drought. The leaves and sheaths of all grasses readily regrow when cut, making grass a perpetual grocery for grazing animals.

IMPORTED FORAGE GRASSES

Man and his herds have followed the grass for ten thousand years, so it is not surprising that European grains and forage grasses soon followed the colonists onto farms in the New World. A common pasture grass, still prevalent on open land all over the East, is Kentucky bluegrass, *Poa pratensis*, actually a native of Europe and Asia. It was first imported to England in the eighteenth century. The American colonists called it "English grass" until about 1720, but as an important range cattle business began to develop in the lower Piedmont states, "English grass" became known as Kentucky bluegrass. *Poa pratensis* now appears spontaneously throughout the East where the humid climate suits it well. Because it forms a dense sod, it is traditionally used for lawns, but this same quality in open meadows tends to inhibit field flowers from seeding.

While successful pasture grasses are able to withstand both trampling and continual cropping, certain other plants thrive in pastures simply because they are unpalatable to livestock. These include the thistles we have already mentioned, as well as buttercups, *Ranunculus acris*, horse nettle, *Solanum carolinense*, a virulently poisonous member of the tomato family, and poke, *Phytolacca americana*. Young sprouts of poke are sometimes eaten like asparagus, but great care must be taken because other parts of the plant are poisonous to humans. Experiments have shown that the seeds of poke must travel through the digestive system of birds or animals before they will germinate. They can then lie dormant in the ground for years waiting for the opportunity to grow. Poke will quickly colonize disturbed soil and often sprouts along paths worn thin by cattle or horses. The berries were once used for dye.

Timothy, *Phleum pratense*, is another European forage grass that when planted in combination with alfalfa makes good horse hay. There are various stories about its arrival here. According to one, it was imported by a New Hampshire farmer named Timothy Hansen in 1731 after he took a trip to France in search of rich grass for his pastures. Somewhat immodestly, he named the new grass for himself

HORSE NETTLE
Solanum carolinense

and immediately began marketing the seeds. Incidentally, the specific name *pratense* or *pratensis* means "of meadows." Not surprisingly, it is attached to many forage plants.

Orchard grass, *Dactylis glomerata*, also came by way of England, where it was called cocksfoot because of its typical three-part ear. Orchard grass grows well in light shade and acquired its American name because farmers planted it in their orchards.

In recent years, highway departments have been responsible for the spread of certain grasses. Fescues, especially the meadow fescue, *Festuca elatior*, were planted along many roads because they grow well in nitrogen-poor soil, while the love grasses, *Eragrostis* spp., were imported from the dry plains of Africa for their drought resistance. Bermuda grass, *Cynodon dactylon*, an alien that is as tough and persistent as crabgrass, was also used along roads.

Native Grasses

Because mesic land in the East was originally covered with woods, we have relatively few indigenous species of grass. Many of these are the useful grasses of marsh and beach that we will examine in a later chapter. Two conspicuous inland species, however, are broomsedge, *Andropogon virginicus*, which often spreads into exhausted fields, and purple top, *Triodia flava*, a waxy, panicled grass noted for its attractive purple ear in late summer and early fall. There are also many species of *Panicum*, or switch grass. One of these, the perennial *P. virgatum*, has become so popular with gardeners that there are now several vegetatively propagated cultivars on the market. There is also a native species of *Eragrostis*, called purple love grass, *E. spectabilis*, that is common on sandy soil, particularly along roadsides. In late summer, the fine-textured flowers appear like a reddish cloud at ground level.

Native grasses are warm season grasses, starting into growth after the soil warms in late spring and spurting upward during the heat of summer. The European forage grasses are cool season grasses. They green up quickly in early spring and flower in early summer, falling dormant during the hot days of July and August. When the cooler days of autumn arrive, they begin again to grow and often stay green all winter.

One cannot talk of native American grass without mentioning the prairie. The first settlers to move west found a vast sea of grass blowing in the wind. It was principally turkey foot, or big bluestem, *Andropogon gerardi*, the backbone of the tall grass prairie. In moist years, big bluestem may grow six feet high. Before the prairie was cut by the plow, a single stand might have been thousands of years old, the tops growing fresh each year from ancient, enduring roots that reached many feet into the earth. The name bluestem comes from a faint blue-green cast to the leaves; the stem is actually tinged with purple. The flowering ear is a three- to five-fingered raceme, resembling a turkey's foot.

Little bluestem, *Schizachyrium scoparium*, described by author Willa Cather as "red as wine stains," grew where the prairie was higher and drier and was often mixed with leadplant, *Amorpha canescens*, and purple prairie clover, *Petalostemum purpureum*, two midwestern members of the pea family.

The third great grass of the tall grass prairie was Indian grass, *Sorghastrum nutans*, a grass that has migrated east in many places. It, too, matures at six feet by the end of summer and has graceful wandlike ears. The foliage turns an attractive yellow-orange in autumn. There is now a cultivar of Indian grass on the market called "Sioux Blue" that was selected at Pennsylvania's Longwood Gardens for its nearly powder blue foliage.

The grass that the early pioneers used to build their sod houses was buffalo grass, *Buchloe dactyloides*. It is the dominant grass of the western Great Plains. Buffalo grass

Evening Primrose
Oenothera biennis

seldom grows higher than five or six inches and forms a tangled mat that resists invading plants. These attributes have made it a popular lawn grass for the drier sections of the country.

Prairie flowers that have spread eastward in large numbers include various species of sunflower, *Helianthus*, as well as black-eyed Susans, *Rudbeckia hirta*, and butterfly weed, *Asclepias tuberosa*. Many others belong to genera that also have eastern species, such as ironweed, *Vernonia*, wild indigo, *Baptisia*, golden ragwort, *Senecio*, and evening primrose, *Oenothera*, as well as the more familiar *Phlox*, gentians, *Gentiana*, *Aster*, and the ever-present goldenrod, *Solidago*.

The prairies are gone now, and many of the fields and pastures of the East have vanished under housing developments and shopping malls. Many others are well on their way into woody succession. Paradoxically, roadsides, where they are not continually mowed or sprayed, are becoming the last refuge of these meadow flowers. In many parts of the country this trend is being actively encouraged in a program called "Operation Wildflower," run by the National Council of State Garden Clubs, Inc. In the Midwest, some groups are now restoring the prairie ecosystem on old fields and worn-out pastures. It is not an easy task: prairie grasses and wildflowers are not normally plants of disturbed soil but part of an ecosystem that originally took thousands of years to evolve. However, it has been shown that if big bluestem can be reestablished on a site, other prairie species often follow.

LEGUMES

Legumes are members of the pea family, Leguminosae, second only to grasses in providing food for man and livestock. A species of bacteria able to convert the nitrogen gas in the atmosphere into compounds usable by green plants lives symbiotically on the roots of leguminous plants such as beans, clover, alfalfa, and peas, enabling them to thrive in nitrogen-poor soils while enriching them in the process. This makes legumes especially valuable to farmers who grow them as livestock feed and use them as "green manure." Legumes include all the commercially grown beans and peas, including soybeans, a plant with so many uses that it was traditionally referred to in Japan as "the little honorable plant." Alfalfa, *Medicago sativa*, red clover, *Trifolium pratense*, and alsike clover, *T. hybridum*, are all legumes that are widely cultivated as forage plants. Alfalfa has the most ancient lineage. It is alluded to in Babylonian texts of 700 B.C., and Pliny recorded that the Persians who invaded Greece in 490 B.C. brought alfalfa to feed their chariot horses.

RED CLOVER
Trifolium pratense

White clover, *T. reptens*, which used to be a common component of lawns, has been largely eliminated by the overuse of broad-leaf weed killers. Unfortunately, its nitrogen-fixing ability has disappeared as well. Other alien but widespread clovers found growing wild in our fields include black medick, *Medicago lupulina*, yellow sweet clover, *Melilotus officinalis*, and bird's foot trefoil, *Lotus corniculatus*. Crown vetch, *Coronilla varia*, is a legume that is widely used along highways.

KUDZU
Pueraria hirsuta

Kudzu, *Pueraria hirsuta*, is an infamous member of the leguminous tribe. During the dust bowl days of the 1930s, kudzu was hailed as a wonder plant that would rejuvenate soil and stabilize erosion on worn-out cotton land in the South, while at the same time providing cheap, fast-growing, and nutritious forage for livestock. Not only that, glowed the promotional literature, it grew up to a foot a day and had the amazing ability to produce a new plant from every node along the stem! Thousands of workers in the Civilian Conservation Corps went south to plant kudzu along highways. Farmers stopped growing cotton and grew kudzu instead. By the end of the Second World War, small southern towns were holding kudzu festivals, a kudzu queen was crowned in Alabama, and Japan and China were both claiming to be the native home of this wonderful plant. Unfortunately, kudzu was soon invading valuable cropland and smothering acres of timber with an impervious curtain of green. Now the problem is how to get rid of it.

Less flamboyant but native legumes include ground nuts, *Apios americana*, wild bean, *Phaseolus polystachios*, milk pea, *Galactia regularis*, and hog peanut, *Amphicarpa bracteata*. All, as their common names suggest, were once used for food. Various vetches, vetchlings, and lespedezas are native legumes, as are the numerous tick trefoils, *Desmodium* spp., that are rarely noticed until the seed pods, known as "beggar's lice", cling to one's clothes in the fall.

Flowers of the pea family are bilaterally symmetrical. They have five petals, but the two lower ones are often joined to form a "keel" that surrounds the stamen and pistil. The two side petals form wings, and the upper one a "banner." Sometimes, as in the clovers, the individual flowers are minute, but are crowded together in a head.

FLOWERS OF MESIC FIELDS

Meadow flowers are dominated by members of the Compositae, or aster family. Composites are the most advanced family of dicots. Dicots, as opposed to monocots, have two seed leaves, flower parts in multiples of two or five, fibrous roots, and net-veined leaves. Each composite flower is really composed of many flowers. A daisy, for instance, is made up of tiny disk flowers, each with its own stamens and pistil, surrounded by ray flowers, again each with stamens and pistil. The tiny individual blossoms can be readily seen with the help of a hand lens. Some composites lack ray flowers, being composed entirely of disk flowers, but all have a structure called an *involucre* beneath the flower that is composed of tiny leaflets, or bracts.

Among the first native composites to bloom is fleabane, *Erigeron annuus*, a prairie flower that spread eastward as the forest was cleared. It has small daisylike flowers with narrow rays and looks rather like a white aster, except that it begins blooming in May and continues for most of the summer. Golden ragwort, *Senecio aureus*, soon follows, another daisylike flower with butter yellow rays and disks. Golden ragwort is followed by drops of yellow dandelions, orange hawkweeds, *Hieracium* spp., and swales of yellow mustards, *Brassica* spp. At their feet may be creeping yellow moneywort, *Lysimachia nummularia*, hiding among the violets in the developing grass. We have dozens of native violets: Most are woodland plants, but the common blue violet, *Viola papilionacea*, is found in open meadows, and careful observation may reveal other species as well. Fragile Quaker ladies, *Houstonia caerulea*, may spread over a grassy bank like tiny lavender dots, while Venus' looking glass, *Specularia perfoliata*, hides among the grasses, too. Its name comes not from the stalkless sky blue flowers but from the flat, polished seeds that follow them.

The first yarrows, *Achillea millefolium*, bloom in June. The flowers of some are not the usual creamy white but a subtle dusky pink. By late June the common milkweed, *Asclepias syriaca*, unfolds heavy pink blossoms that hang like fragrant popcorn balls in the leaf axils. Milkweeds have distinctive, intricate flowers. The five petals are attached to a clublike column and hang down rather like a tiny hula skirt, hiding the sepals. The stamens and pistil are concealed behind "hoods" at the top of the column, and each hood ends in a peculiar hornlike structure. This strange arrangement seems to discourage any insects but bees and butterflies from pollinating the flowers. Only their strong legs can bend down the hoods and reach the pollen.

GOLDEN RAGWORT
Senecio aureus

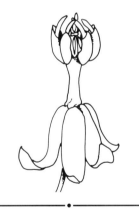

MILKWEED FLOWER
(ENLARGED)
Asclepias spp.

∴ FIELD AND FOREST ∴

If you are lucky, you may find a pasture rose, *Rosa carolina*, blooming in your meadow. It is a shy rose with single shell pink flowers, nowhere near as aggressive as the Asian invader, *Rosa multiflora*. Soon, the slender blue-eyed grasses, *Sisyrinchium* spp., dot the true grass with tiny blue six-point stars. They are members of the iris family, Iridaceae. Looking at first glance like a pink species is the alien Deptford pink, *Dianthus armeria*, a diminutive member of the carnation family, Caryophyllaceae.

By July first, the meadow belongs to the robust black-eyed Susans, *Rudbeckia hirta*, a prairie flower that has moved east in abundance. Mixed with St. John's wort, *Hypericum* spp., they stain the summer meadow gold, accented here and there with the pale yellow of the common evening primrose, *Oenothera biennis*. Butter and eggs, *Linaria vulgaris*, blooms like tiny orange and yellow snapdragons, while chicory, *Cichorium intybus*, and Queen Anne's lace, *Daucus carota*, decorate the roadsides with blue satin and eyelet.

By August early goldenrod appears, *Solidago juncea*, and all the weedy plants with homely names reminiscent of vanished barnyards: lambsquarters, *Chenopodium album*, goosefoot, *C. hybridum*, and sow thistle, *Sonchus* spp. The air is filled with the hot, dusty smell of ripening corn, and Turk's-cap lilies, *Lilium superbum*, nod along the stone walls of New England like opulent visitors from the Orient. Each exotic recurved flower has a green star in its throat and conspicuous stamens like small brown polo mallets. Lilies seem far too tropical for these northern climes, but in fact, some twenty species are native to North America.

In early September the goldenrods and asters color the landscape yellow and mauve. There are seventy-five species of goldenrod native to the East, and they vary considerably in height, habit, and choice of habitat. The two most common in old fields and pastures are the tall goldenrod, *Solidago altissima*, and the rough-stemmed goldenrod, *S. rugosa*.

TURK'S-CAP LILY
Lilium superbum

They are easily distinguished by their leaves. Tall goldenrod has narrow leaves with three parallel veins, while the leaves of rough-stemmed goldenrod are feather veined. Both have stout, hairy stems, although tall goldenrod's is more grayish and downy. A third common species is Canada goldenrod, *S. canadensis*, easily identified because it is the only species attacked by a midge larvae that causes crown gall, a conspicuous proliferation of leaves at the apex of the stem. The spherical swelling farther down the stem found on several other species is caused by the larva of a fly. The fly lays its eggs on the stem. When the larva hatch and begin to feed on the plant's tissue, it reacts by encasing them in a gall.

New England Aster
Aster novae-angliae

New England asters, *Aster novae-angliae*, bloom on moist or mesic fields, forming stout clumps of royal purple daisies as beautiful as any garden perennial. In fact, most garden asters are selections of this species; nurserymen collect seedlings that vary from the standard color and then propagate them vegetatively. The last plants to bloom before winter are the white asters or frostweeds, *Aster ericoides*, *A. pilosus*, *A. simplex*, and others. They are stout, bushy plants with myriads of tiny white daisies, called frostweeds because they go on blooming long after everything else has finished. The yellow centers to the flowers usually turn reddish brown after the plant is fertilized, but in at least one species, the calico aster, *A. lateriflorus*, they turn a distinctive purple.

Flowers of Dry (Xeric) Fields

Dry meadow plants are found on old fields that have lost topsoil to erosion, along the shoulders of graded roads, or

BIRD'S-FOOT VIOLET
Viola pedata

PRICKLY PEAR CACTUS
Opuntia humifusa

PEARLY EVERLASTING
Anaphalis margaritacea

on sandy soil near the coast. Imported forage grasses do not grow well on dry soils, so on dry hillsides they are quickly replaced by broomsedge, *Andropogon virginicus*. In especially dry places, poverty grass, *Aristida dichotoma*, is common as well.

Our most beautiful native violet, the bird's-foot violet, *Viola pedata*, blooms in May on sterile, sandy soil. It has large, almost pansylike flowers of a vivid blue-violet and finely cut leaves rather like those of an anemone. Bird's-foot violets take up to three years to bloom from seed, although once established the plants may persist for a decade. Prickly pear cactus, *Opuntia humifusa*, is the only true cactus that is widespread in the East. It has jointed flat pads, which are actually modified stems, and tufts of bristles. A related species of prickly pear was once imported to Australia as a living fence to segregate cattle and went wild over the landscape in the same way kudzu has here. Other dry meadow forbs include the pussytoes of the genus *Antennaria*, which unfold their silvery "paws" in June. Their cousins the everlastings come on in August and September. The most showy of these is the perennial pearly everlasting, *Anaphalis margaritacea*, which may form large colonies in sandy soil. The annual cudweed, *Gnaphalium obtusifolium*, is less showy but has a delicious spicy scent of curry.

Eupatorium hyssopifolium is a dry meadow species of boneset and *Solidago nemoralis* a dry-loving goldenrod that is a compact, almost dwarf plant with handsome gray foliage. The lavender bergamot, *Monarda fistulosa*, and the blazing stars, *Liatris* spp., are prairie plants that have spread east to dry meadows in some places. Butterfly weed, *Asclepias tuberosa*, the handsome orange milkweed of dry hillsides and old railroad embankments, and the yellow coreopsis, *Coreopsis lanceolata*, are also flowers of the prairie.

Goatsbeard, *Tragopogon pratensis*, looks like a huge flat-topped dandelion with square petals and has a seedhead like a three-inch pom-pom. Ground cherry, *Physalis* spp., is a dry-loving member of the tomato family, Solanaceae, that bears sweet yellow fruit concealed within a papery husk. But beware you don't confuse it with horse nettle, *Solanum carolinense*, another member of the same family, which is deadly poisonous! Oddly enough, horse nettle was once given in minute doses as a cure for tetanus.

Flowering spurge, *Euphorbia corollata*, is another prairie plant that grows in the East on dry sandy soils. It is a smooth, almost succulent plant with showy white flowers rising from a whorl of smooth green leaves. What appears to be the petals of the flower are really bracts that surround the tiny true flowers in the center. It blooms in August. Two of our most beautiful asters bloom on sandy soils: the showy aster, *A. spectabilis*, and the stiff aster, *A. linariifolius*, a dwarf plant with leaves as stiff as spruce needles. Both have lovely lavender flowers and are well worth cultivating in the garden.

FLOWERING SPURGE
Euphorbia corollata

FLOWERS OF WET (HYDRIC) MEADOWS

Wet meadow species congregate along the course of a stream or in a poorly drained swale or roadside ditch. They may also be clustered at the bottom of a hill, where the soil is rich and moist, perhaps even in the same field as some of the upland species, although the two groups never intermix.

In April and May, the golden alexander, *Zizia aurea*, blooms along brooks and in low meadows. It looks rather like a yellow Queen Anne's lace and is, in fact, a member of the same family, Umbelliferae, the parsley family. Two other yellow umbelled flowers that may be confused with it are the meadow parsnip, *Thaspium trifoliatum*, and the alien

VIRGINIA BLUEBELL
BRANDYWINE BLUEBELL
Mertensia virginica

GREEN-HEADED
CONEFLOWER
Rudbeckia laciniata

wild parsnip, *Pastinaca sativa*. The central cluster of flowers on *Zizia* are stalkless, while those of *Thaspium* are stalked, a distinction easily made with a hand lens. *Pastinaca* is never found in wet soils. It is a stout upland species that spreads over dry hillsides. Two native white mustards and one invasive alien weed bloom on damp ground in early spring. The natives are Pennsylvania bittercress, *Cardamine pensylvanica*, and spring cress, *C. bulbosa*. The weed is garlic mustard, *Alliaria officinalis*. All are members of the family Cruciferae. The most beautiful flower of streamsides and floodplains, however, is the Virginia bluebell, *Mertensia virginica*. Its clusters of nodding pink buds open to pure blue bell-like flowers in May.

In June the fringed loosestrife, *Lysimachia ciliata*, blooms. Its single yellow flowers, each with tattered petals, nod from the leaf axils. Nearby, elderberries, *Sambucus canadensis*, and the gray-stemmed dogwood, *Cornus amomum*, may bloom with a swamp rose, *Rosa palustris*. Swamp rose is larger than the upland pasture rose and has flowers of a deeper pink. Its thorns form tiny hooks, while those of the pasture rose are straight.

As June moves into July, the green-headed coneflower, *Rudbeckia laciniata*, appears on floodplains and riverbanks, while spotted jewelweed, *Impatiens capensis*, and its yellow sister species, *I. pallida*, begin to show color along every brook and watercourse. The sepals of each jewelweed flower form a spur full of sweet nectar that attracts bumblebees, hawkmoths, and hummingbirds. Jewelweeds are annuals that have evolved many tricks to ensure the healthy survival of the species. For instance, each flower produces ripe pollen before it exposes a stigma, thus neatly avoiding self-fertilization that might gradually weaken the species. However, in the event of a year so dry it limits flowering altogether, the plants produce special self-fertile flowers, as a sort of emergency measure to ensure seed production. But the jewelweed's system of seed dispersal is perhaps the most spectacular of all its talents. The merest touch of a ripe pod causes it to spring open with a snap, flinging the seed in all directions. No wonder children call them "touch-me-nots."

One of our most beautiful native lilies blooms on open, moist ground in July. This is the meadow, or Canada lily, *Lilium canadense*. It grows over six feet tall, dangling waxy, bell-shaped flowers from long stems that rise from a whorl of smooth leaves. Unfortunately, it is increasingly rare. July is also the month that the dusky pink spires of hardhack, *Spiraea tomentosa*, appear, often mingled with swamp milkweed, *Asclepias incarnata*. In this species the typical milkweed flowers are collected in a flat-topped inflorescence of iridescent pink.

TOUCH-ME-NOT
JEWELWEED
Impatiens capensis

By August, the Joe Pye weeds begin: their domes of pink froth rise on purple-spotted stems above a whorl of rough textured leaves. Our largest, most stately Joe Pye is the hollow-stemmed *Eupatorium fistulosum*, which towers a sturdy six feet or more. Joe Pye is believed to have been the name of a native American herbalist who lived near Salem, Massachusetts and used this plant as a cure for fevers.

As summer begins to wane, the spires of cardinal flower, *Lobelia cardinalis*, glow like red velvet in low meadows and pondsides. Its sister species, the blue lobelia, *Lobelia siphilitica*, blooms among the creamy bonesets and yellow lance-leaved goldenrod, *Solidago graminifolia*. The bold New York ironweed, *Vernonia noveboracensis*, holds up shaggy flowers of a deep red-violet. The color is a perfect foil for the pale lavender of the purple-stemmed aster, *Aster puniceus*, the most abundant species along streams and in low meadows of the Piedmont. Mixed among it one might find monkey flower, *Mimulus ringens*, or turtlehead, *Chelone glabra*. If you squeeze the flower, the "turtle" will open its mouth.

August and September are also the time for the sunflowers, *Helianthus* spp., their cousins the sneezeweeds, *Helenium* spp., and the tickseed sunflowers, *Bidens* spp. When

JERUSALEM ARTICHOKE
Helianthus tuberosus

SNEEZEWEED
Helenium autumnale

most people think of sunflowers, they think of the common sunflower, *Helianthus annuus*, that is the parent of all the large-disked giants grown in gardens. Native populations used the seeds of this species for both food and oil, and this is thought to be the reason it is so widespread in the Midwest. Now it has come east as well and oddly enough grows abundantly near the Philadelphia Airport, raising its circular yellow heads from every drainage ditch, usually out of a bed of beer cans, and often in concert with the brilliant spires of purple loosestrife, *Lythrum salicaria*, a showy but dangerously invasive immigrant from Europe. A more wide-ranging sunflower in the East is the Jerusalem artichoke, *H. tuberosus*. The name Jerusalem artichoke supposedly comes from the Italian word for sunflower, *girasole*. It is no relation of the artichoke, but its root is an edible rhizome that tastes rather like water chestnuts. It is sometimes sold in supermarkets under the name of "sunchoke."

Tall sunflower is another species often found in low meadows in the East. It is aptly named *H. giganteus* for it sometimes grows to ten feet or more, towering over the surrounding vegetation. Tall sunflower has stout red stems and rough, lance-shaped leaves.

Sneezeweeds, *Helenium* spp., differ from the true sunflowers by their buttonlike disks and scalloped petals. We have two species, *H. autumnale*, and *H. nudiflorum*, both of which grow in wet meadows. The name sneezeweed is a false charge. No showy flower designed to be pollinated by insects is responsible for allergies. This is the fault of wind-pollinated plants that bloom at the same time, such as ragweeds and grasses.

The genus *Bidens* includes several showy species as well as some that go unnoticed until they produce the two-pronged seeds called "beggar's ticks" or "Spanish needles." One of the showiest is the tickseed sunflower, *Bidens polylepis*, a midwestern species that has invaded the mid-Atlantic states in abundance, especially on the Coastal Plain, where it turns roadside ditches and low-lying fields into a sheet of brilliant yellow. It is an annual that self-seeds abundantly.

Last in the meadow to bloom are the gentians. The genus is named for a King Gentius who ruled Illyria (now Yugoslavia) about two thousand years ago, and who supposedly used the roots for medicine. With one exception, all eastern gentians are a deep, clear blue. Some, such as the bottle gentian, *Gentiana andrewsii*, and the closed gentian, *G. clausa*, have self-fertile flowers that never open. The most famous species, the fringed gentian, *G. crinita*, is now very rare. It is a true biennial that takes two years to bloom and then dies. If a field where fringed gentians grow is mowed at any time during the summer, the flowering stalks will be cut and the plants will die without setting seed. Studies have shown that fringed gentians also need a high level of magnesium in the soil and a neutral pH of 7.0. So particular are their requirements, in fact, that it is surprising to read how plentiful they were in eastern meadows around the turn of the century.

The many species of sedges and rushes are also reliable indicators of damp soil and low meadows. We will look at them in more detail in the next chapter, when we move closer to the water and examine wetlands.

FRINGED GENTIAN
Gentia crinita

·S·E·V·E·N·

Wetlands

Scant oxygen is the critical environmental factor for wetland plants. Most oxygen is, of course, produced by plants as a by-product of photosynthesis, but oxygen is also used by plants for respiration, just as it is by all living organisms except yeasts and certain bacteria. In its most basic sense, the term *respiration* means the release of energy from food by oxidation: energy that is necessary for growth and all the complex chemical reactions that sustain life.

The roots of land plants absorb oxygen from tiny air spaces found between the particles of a well-drained soil. In saturated soil these spaces are filled with water. However, because the amount of oxygen that can be dissolved in water is less than 5 percent of that which would be available in an equivalent volume of air, aquatic plants have had to develop special tissues that can store air as well as keep them afloat. Many also have a hard coating on their foliage much like that which protects desert plants from desiccation, except in this case the purpose is to keep air in and water out.

All terrestrial plants originally migrated onto dry land from a water environment. Aquatic plants then returned, presumably to escape competition. They did not simply stay behind. We know this because they hold their flowers above the surface to be pollinated and subsequently develop seeds. It was the evolution of seeds, with their unique capacity to store offspring potential safely sealed in a weatherproof case, that enabled plants to leave the water and colonize dry land in the first place.

The water in aquatic habitats may be fresh, brackish, or salt, with pH factors that range all over the scale. Many freshwater plants can tolerate some degree of brackishness, but those of the true salt marsh are a separate community. Peat bog communities are distinctive, too, because bog water is too acid to support the bacteria that decay organic matter, severely limiting available nutrients.

PONDS AND SWAMPS

Natural ponds occur in valleys when water drains from surrounding hills to form sheets of standing water, or a river's current severs an "oxbow" from its own meandering course. Beavers and landslides often dam high mountain streams to create shallow ponds studded with the pale trunks of drowned trees. Yet most olive-brown country ponds that are edged with cattails and smeared with lily pads are artificial, dug to water cattle or support fish and waterfowl. Often a low spot by a woodland stream may be all that is left of such a pond, constructed over a century ago to provide power for a grist- or sawmill.

Every pond, whether natural or artificial, is a successional habitat. All ponds eventually grow old and die, successively becoming cattail marshes, shrub swamps, wooded swamps, and eventually dry woods. The process, called *eutrophication*, is natural, yet today it is often unnaturally accelerated by runoff from cultivated and developed land as well as by waste water containing phosphates that fertilize and stimulate the growth of water plants.

Lakes are successional, too, although a lake may take thousands of years to fill. While there are still glacier lakes today that are far too deep and cold for plants except at the edge, others have been transformed over thousands of years into bogs and prairie potholes.

It is plants that age a pond. Advancing into the center in concentric rings, they trap silt and gradually fill the bottom with organic debris. As the pond becomes shallower, each group of plants creeps farther into its neighbors' habitat until the open water disappears altogether. This process begins immediately upon the birth of a pond, as thousands of seeds blow in on the wind or float down a feeder stream while hundreds more hitch rides on the muddy feet of wading birds.

MARSH MARIGOLD
Caltha palustris

A pond in its productive prime is called mesotrophic. We might approach such a pond through a wooded swamp, once a pond itself, but now only flooded intermittently by runoff. The water level in such a swamp may be as deep as a foot in winter and spring, yet dry up completely during a late summer drought. If the swamp is old enough, it may be well on its way to becoming a woods filled with silver or red maple, *Acer saccharinum* or *A. rubrum*, and sycamore, *Platanus occidentalis*, trees also common on floodplains where they are subject to periodic flooding. Or it may become a thicket of black willow, *Salix nigra*, alder, *Alnus rugosa*, and elderberry, *Sambucus canadensis*, perhaps sparked in winter by the brilliant red berries of the winterberry, *Ilex verticillata*, or the peculiar globular seed pods of buttonbush, *Cephalanthus occidentalis*. If the season is early spring, marsh marigolds, *Caltha palustris*, may glow like glistening yellow buttercups in mud dotted with the purple hoods of skunk cabbage, *Symplocarpus foetidus*, while in summer the air may be sweet with the fragrance of the coastal sweet pepperbush, *Clethra alnifolia*, or swamp azalea, *Rhododendron viscosum*. Swamp azalea was the very first American plant to travel back to England; seeds were sent to England by the Virginia colonists in 1680. In summer, too, wet meadow species described in the last chapter will be found on the banks of a pond.

SWAMP AZALEA
Rhododendron viscosum

∴ WETLANDS ∴

EMERGENTS

The shallow water close to the bank of the pond is known as the littoral zone. This is the domain of the emergents, plants with their roots anchored firmly in submerged soil, but with stems, leaves, and flowers held well above the surface. An example is the broad-leaf cattail, *Typha latifolia*. Cattails are monoecious: they have separate male and female flowers on the same plant. In this case, the male flowers are clustered at the top of the stem and fall away as soon as the female flowers that form the familiar brown cigar-shaped inflorescence are fertilized.

Cattails often grow in the littoral zone with two members of the arum family, Araceae: golden club, *Orontium aquaticum*, and sweet flag, *Acorus calamus*. The individual flowers of golden club are minute, but they are arranged on a conspicuous yellow-tipped "club," botanically called a spadix. Skunk cabbage and jack-in-the-pulpit are also members of this family and have similar flowers, but their spadixes are concealed by a hood. The folk name for golden club is "never wet" because its flat leaves have a waxy coating that sheds water like a duck. Sweet flag, on the other hand, has rigid swordlike leaves, fragrant when crushed, and a spadix of tiny, tightly packed flowers that jut out at an angle from the stem. Sweet flag was called *wisuhkum* by the Lenni-Lenapes, a word meaning "it tastes bitter." They mashed the root and used the juice to treat heart disease.

Blue flag, *Iris versicolor*, has similar leaves yet is a very different plant. This lovely native iris blooms in shallow water in May, just before its robust alien cousin, the handsome yellow *I. pseudacorus*. Swamp loosestrife, *Decodon verticillatus*, also known as water willow, is a summer-blooming emergent with tufts of lavender flowers in the upper leaf axils. Water willow is a shrubby plant with arching branches that root at the tips. It forms dense thickets that trap decaying vegetable matter and gradually extend the

CATTAIL
Typha latifolia

BLUE FLAG
Iris versicolor

ARROWHEAD
Sagittaria latifolia

PICKERELWEED
Pontederia cordata

UMBRELLA SEDGE
Cyperus strigosus

shoreline of the pond. Water willow belongs to the same family, Lythraceae, as the alien purple loosestrife, *Lythrum salicaria*. Other plants that are often called loosestrifes, such as fringed loosestrife, *Lysimachia ciliata*, and swamp candles, *L. terrestris*, are actually members of the primrose family, Primulaceae.

Arrowhead, *Sagittaria latifolia*, is another common plant of the littoral zone. Native Americans called it "duck potato" for its edible root, a golf-ball-sized tuber that ducks do, in fact, pull up and eat. The pale blue spires of pickerelweed, *Pontederia cordata*, may form large masses in shallow water at the edge of ponds, while various sedges and rushes are reflected in jagged stripes across the quiet surface.

Wool grass, *Scirpus cyperinus*, a conspicuous grasslike plant with a floppy brown inflorescence, is not a grass but a sedge. Umbrella sedge, *Cyperus strigosus*, is another species found at the edge of ponds. Most sedges have stems that are triangular in cross section ("sedges have edges"), have narrow leaves arranged in threes, and prefer wet places. They are common not only near ponds, but in freshwater marshes, roadside ditches, and even in poorly drained fields. The river of grass that is the Florida Everglades is actually a river of sedge, *Claudium jamaicensis*.

Some sedges are popularly called bulrushes, genus *Scirpus*, or spike rushes, genus *Eleocharis*, a practice that thoroughly confuses the issue. True rushes belong to the family Juncaceae and have stems as round as knitting needles. Two rushes commonly found on the edges of ponds are soft rush, *Juncus effusus*, and Canada rush, *J. canadensis*. The so-called chairmaker's rush is not a rush but a sedge, *Scirpus americanus*. It has a sharply triangular stem with a scaly, brown flower cluster halfway up the side.

Two grasses that may form large patches in wet, marshy spots are reed canary grass, *Phalaris arundinacea*, and blue joint, *Calamagrostis canadensis*. Reed canary grass makes excellent hay and is often planted on more upland sites.

Floating Leaved Plants

Beyond the emergents of the littoral zone, but still in relatively shallow water, we come to the floating leaved plants. They have long pliable stems and round padlike leaves with no lobes or serrations that might be torn by wind or waves. The leaves are often protected by a hard, waxy coating to prevent waterlogging as well, for the water that buoys them up also thrusts their flat surfaces into the full force of pelting rain and hail.

Floating leaved plants are true aquatics and include the waxy, fragrant waterlily, *Nymphaea odorata*, as well as the coarser, more invasive pond lily or spatterdock, *Nuphar variegatum* or *N. advena*. Water shamrock, *Marsilea quadrifolia*, is a floating leaved fern that looks rather like a large four-leaf clover. It has spread throughout the East from a pond in Litchfield County, Connecticut, where it was introduced in 1862.

American lotus, *Nelumbo lutea*, is a beautiful native aquatic that is not common in the East, although it is widespread on parts of the Mississippi River. Lotus flowers are a clear yellow, and the seed pod that follows them looks like a perforated brown funnel. It is often sold by florists for dried arrangements. The round leaves of the lotus are peltate, meaning that the stalk is attached to the center, rather like an inverted umbrella.

Floating hearts, *Nymphoides* spp., are delicate aquatic members of the gentian family, Gentianaceae. They have small, waterlilylike leaves and umbels of tiny, white, five-petaled flowers. *N. aquatica* is found on the Coastal Plain south of New Jersey. *N. cordata* is a smaller, more northern species.

Spatterdock
Nuphar advena

∴ FIELD AND FOREST ∴

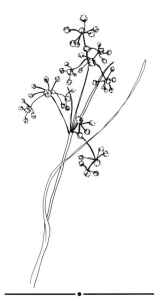

CANADA RUSH
Juncus canadensis

FREE-FLOATING AQUATICS

Free-floating aquatics have no contact with the bottom, but float in tangled masses on the surface of the water where they are vulnerable to pounding by waves and swift currents. For this reason they are only found in still water, where they may form mats dense enough to clog navigable waterways. Most floating aquatics grow in a form of a rosette and multiply by means of long stems called rhizomes. Their leaves and stems are often swollen with air tissue and act like bladders to keep them afloat, while their hairy roots dangle beneath, absorbing nutrients from the water. Duckweed, *Spirodela polyrhiza*, is a common floating plant of eastern ponds. Its tiny, swollen "leaves" are not leaves at all, but modified stems. Duckweed may blanket a pond in summer, filling the space between the waterlilies like coarse green algae. In winter, it simply drops to the bottom, hiding beneath the ice until the warm water of spring causes it to float to the surface again.

Water hyacinth, *Eichhornia crassipes*, is a floating aquatic with beautiful blue flowers and glossy, deep green foliage, but it is a notorious weed of Coastal Plain waterways south of Virginia. We know that specimen plants of water hyacinth, a native of South America, were given as souvenirs at an International Cotton Exposition held in New Orleans in 1884. They were soon clogging river channels all over the South. Happily, it has been discovered that water hyacinths can purify polluted water, so the plant is now being put to good use in sewage and factory-waste lagoons.

FRESHWATER TIDAL MARSHES

All the plants discussed so far grow in nontidal waters of inland ponds and rivers, yet the influence of the tide can also be felt along the margins of coastal rivers and bays many miles from the sea. A conspicuous example is the Chesapeake Bay and its tributaries, where a wedge of brackish water develops as the heavier salt water sweeping in on the tide sinks beneath fresh water flowing down the rivers. Not only does this brackish wedge move about on the tide, it also travels up- or downstream from season to season according to the amount of rainfall. Thus, the marshes that develop in coves along estuarine rivers may grow in water that is sometimes fresh, sometimes brackish. Many pond plants, such as arrowhead, cattails, and pickerelweed, can tolerate such minor fluctuations of salt content and often grow in river marshes. Soft rush, *Juncus effusus*, is found here as well, often growing with arrow arum, *Peltandra virginica*, a plant with the distinctive spadix of the arum family and large, triangular leaves. The smartweeds, genus *Polygonum*, also thrive here. Three species often found are *P. punctatum*, *P. densiflorum*, and *P. hydropiperoides*. All have tiny pink or white flowers that are clustered in a terminal spike. Characteristic of smartweeds is a tissuelike sheath that surrounds the nodes on the stem. Halberd-leaved "tear thumb," *P. arifolium*, is a pesky member of the same genus. It has sharp, downward-pointing barbs all along its stem and drapes itself over other marsh plants like a spiny fishnet. Northern wild rice, *Zizania aquatica*, is a valuable grass of estuarine marshes, with grains that are a gourmet treat for both man and waterfowl. The plant is distinctive because the male flowers rise on a plumelike inflorescence at the top of the stem while the female flowers fan out beneath.

ARROW ARUM
Peltandra virginica

In late summer, rose mallows decorate these marshes with enormous pink or white flowers that look as if they were fashioned from crepe paper. *Hibiscus moscheutos* may be pink or white, but each blossom has a center of deep carmine. The pure pink *H. palustris* is more common in New England. Rose mallows are sturdy plants with large, heart-shaped leaves. They have sculptured buds and pointed bracts that surround the developing seed pods like a cage. Rose mallows make handsome garden plants; they do not seem to require sodden soil and can be easily grown from seed.

Sometimes the frosty pink heads of *Mikania scandens* twine among the mallows in the marsh. It is the only climbing member of the Composite family native to North America with foliage rather like a morning glory.

As fall approaches and the rose mallows fade, two species of tickseed sunflower paint some river marshes a brilliant yellow. They are *Bidens coronata*, which has rather delicate, deeply lobed leaves, and *B. laevis*, with thick, almost succulent, opposite leaves. Towering twelve or more feet above them may be a single stately plant of narrow-leaved sunflower, *Helianthus angustifolius*.

Many river marshes, particularly where the natural water level has been artificially altered, have been invaded by *Phragmites australis*, the giant reed. A member of the grass family, Gramineae, phragmites can form impenetrable stands, to the exclusion of all other plants. It has become an unwelcome symbol of the urban Northeast, appearing in roadside ditches, dumps, industrial parks, and wherever the drainage is poor. A conspicuous example is the so-called Jersey Meadows, a marshy plain covering thousands of acres just across the Hudson River from New York City. It is sad to think that these meadows were once covered with a variety of freshwater marsh plants and, according to an early New York botanist named John Torrey, included a white cedar swamp full of rare orchids. "No region is more interesting to a botanist . . . than that which surrounds the City of New York," he wrote in 1819. The reeds apparently began to invade in 1867 when dikes were built that changed the level and flow of water.

Viewed objectively, phragmites is a handsome plant, with large terminal panicles of silky purple flowers that turn

Rose Mallow
Hibiscus palustris

Climbing Hemp
Mikania scandens

gray in the fall and winter. Unlike most weeds of modern civilization, it is not an alien, but a plant of worldwide distribution. Remains preserved in ancient peat bogs show it has been on this continent for thousands of years, and there is evidence that the native populations used it for arrow shafts well before the coming of the Europeans.

Phragmites needs a wet soil with a neutral pH, so its rapid spread often indicates that sewage pollution, which is alkaline, has altered the natural acidity of the water. Many stands are holdovers from the days when dredge spoil from ditches and canals was commonly dumped onto salt marshes, raising the original level of the land. The reeds cannot tolerate fluctuating water levels such as those that occur with the natural rise and fall of the tide, so the key to eliminating them seems to lie squarely in intelligent land management. Mowing or burning merely encourages stronger growth, and herbicides work only temporarily. Most herbicides cannot be used near watercourses in any case, for they have serious side effects.

SALT MARSHES

A true salt marsh is a fluid plain of meandering creeks, black mud, and grass—a transitional zone between land and sea. Salt marshes typically develop behind barrier islands where silt pushed down rivers mixes with sand washed inland by the sea. These sediments are quickly invaded by salt-tolerant plants, many of which have cordlike roots that sift and bind them, gradually extending the land. Extensive salt marshes are found on the Coastal Plain, where a small rise in sea level will flood the low-lying land with salt water as far as the tide can reach. New England marshes are found on the coastal outwash plain, those heaps of sand and gravel that were left behind by melting glaciers.

Few species of plants can grow in the hostile environment of a salt marsh. Paradoxically, although it is wet much of the time, a salt marsh is a physiologically dry habitat for plants, because salt concentrations in the surrounding water cause living cells to lose water by osmosis. Without special adaptations, plants in such an environment would shrivel and die.

For instance, many species of *Spartina*, the principal genus of grass on the salt marsh, have developed a membrane on their roots that excludes salt, while at the same time cells in other parts of the plant actually allow the salt to concentrate in order to counteract osmotic pressure. *Spartina* even has glands on its leaves that excrete salt. Yet in spite of all this, experiments have shown that not only does *Spartina* grow luxuriously in fresh water, its seeds will only germinate in fresh water, indicating that the plants must have moved onto the salt marsh in order to escape competition from more robust freshwater plants.

SALT HAY
Spartina patens

There are three species of *Spartina* found in eastern salt marshes, and each is dealt a particular habitat by the tide. At the seaward edge and along the fingerlike creeks that invade the marsh, where the mud is black, sticky, and constantly flooded, we find the saltwater cordgrass, *S. alterniflora*. It is the colonizer of the marsh, with strong underground rhizomes that trap silt and extend the land. On the southern Coastal Plain, where large areas of marsh are flooded daily by the tide, this is the dominant grass. On the higher salt marshes north of Virginia, the salt hay, *S. patens*, colonizes that part of the marsh that is only reached by spring tides that occur with the full and new moon. The third species of *Spartina*, *S. cynosuroides*, grows along the marsh's landward boundary where the water is less salty, safely out of reach by all but the highest storm tides. *S. cynosuroides* is a tall, handsome grass that turns a pale ocher in fall and winter. It is not found north of Connecticut.

Spartina cynosuroides

The *S. patens* of the high marsh has a tidy, mowed look

and is usually dry enough to walk upon; in fact, cattle were grazed here in earlier days and the grass was frequently mowed for fodder. Spike grass, *Distichlis spicata*, mixes with it in more waterlogged areas, and interwoven with both are drifts of the delicate black rush, *Juncus gerardi*, replaced farther south by the species *J. roemerianus*. All are short, slender plants, and their varying colors blend and mix in a subtle abstract design, especially beautiful in autumn when they turn yellow ocher, burnt sienna, and umber.

In among the grasses are bare areas called salt pannes, where evaporation has apparently caused the salt to accumulate in high concentrations. Here, only *Salicornia* will grow. It is a true halophyte, or salt-seeking plant, that will not grow in a freshwater environment. *Salicornia* has cylindrical fleshy stems, green in summer, carmine in autumn. It is a member of the goosefoot family, Chenopodiaceae, and, like most other members of that family, is edible. Country people used these plants to give a salty tang to salads and pickles and called them "chicken toes," "glasswort," or "saltwort."

Bayberry
Myrica pensylvanica

No woody plants grow on the open marshes of the East Coast as mangroves do in Florida, and some botanists think that is because the ice that glazes the marsh in winter severs woody seedlings. However, three woody shrubs do grow along the landward edge, colonizing the intermediate zone between marsh and dry land where they are only touched by the highest storm tides. They also cluster on high spots in the open marsh forming dark islands in a sea of grass. Both bayberry, *Myrica pensylvanica*, and its more southern relative, sweet gale, *M. cerifera*, grow here. So does the marsh elder, *Iva frutescens*, an awkward shrub with the typical fleshy leaves of plants growing in a maritime environment. The high tide bush, *Baccharis halimifolia*, is a woody member of the composite family, inconspicuous until early fall when the developing seed heads look like hundreds of tiny white plumes. American holly, *Ilex opaca*, fills the

Marsh Elder
Iva frutescens

HIGH TIDE BUSH
Baccharis halimifolia

SALT MARSH FLEABANE
Pluchea purpurascens

SALT MARSH ASTER
Aster tenuifolius

∴ FIELD AND FOREST ∴

woods that grow along the borders of salt marshes on the Coastal Plain, often mixed with two relatives, inkberry, *I. glabra*, a shining evergreen shrub that is much used in the nursery trade, and the brilliant red winterberry, *I. verticillata*.

Flowers of the salt marsh meadow are few. Most do not appear until late summer or fall. One of the earliest to bloom is salt marsh milkweed, *Asclepias lanceolata*. It is not common but may sometimes be found growing along the narrow rim where marsh and upland meet south of New Jersey. Salt marsh milkweed has flowers of a brilliant orange, reminiscent of butterfly weed. The plant is slender and more delicate than its inland cousin, but the individual flowers are large. Another choice but uncommon plant of the salt marsh is the fenrose, *Kosteletzkya virginica*, a delicate pink member of the mallow family, Malvaceae. With its characteristic bushy column of fused stamens and pistil, it looks like a smaller version of the robust rose mallow of brackish marshes. It is the sole northern representative of a large tropical genus that is named after a Czech botanist, B. F. Kosteletzky, and is not found north of Long Island.

By midsummer, the tiny pink seaside gerardia, *Agalinus maritima*, hides its pink bells among the salt hay, often unnoticed until nearly trod upon. Sea lavender, *Limonium nashii*, has sprays of tiny lavender flowers, each only about an eighth of an inch in diameter, but so numerous as to give the entire marsh a purple haze in late summer. Salt marsh fleabane, *Pluchea purpurascens*, is bolder, forming fuzzy pink masses along the creeks that lace the marsh. In September, salt marsh aster, *Aster tenuifolius*, appears. It is a slender, succulent plant with smooth, narrow leaves and delicate white or lavender blossoms.

Late fall brings the goldenrods. Two that grow in salt marshes are the succulent and showy seaside goldenrod, *Solidago sempervirens*, and Elliot's goldenrod, *S. elliotii*, a slender plant with smooth, elliptical leaves that hug the stem.

Bogs

Streams bubbling into a pond, tides gliding over a salt marsh, river currents sweeping seaward—all renew the dissolved oxygen supply and bring fresh silt rich in nutrients for use by wetland plants. A bog, however, has no flowing water, very little oxygen, and no silt. Bogs—sterile, soggy, acidic, and characterized by quaking mats of vegetation—are the typical wetland community of northern latitudes. Here, where the cool climate inhibits evaporation, bogs spread like watery mantles over shallow depressions originally left by the glaciers. These depressions are all that remain of ancient lakes that have slowly filled with peat over thousands of years.

The podzol soils of northern forests are typically sandy and too acid to readily form humus. In fact, sandy soils, wherever they occur, tend to be deficient in soil bacteria and thus conducive to bog formation in areas of poor drainage, particularly near the coast where fog cools the summer climate. In southern New England, Cape Cod, and offshore islands, such as Martha's Vineyard and Nantucket, where the gravelly soil was originally deposited by glaciers, such bogs develop in depressions called "kettle holes."

South of the boundary that once separated glaciated land from unglaciated, bogs are less common, yet they do occur in poorly drained depressions in the Appalachians where complicated geologic formations have resulted in impermeable soils and perched water tables. They also occur in the sandy soil of the New Jersey Pine Barrens where the aquifers are close to the surface. All these bogs contain northern plants that were pushed southward by the glaciers.

Carolina bays are shallow ponds on the sandy Coastal Plain where the soil is acid and the drainage poor. They develop bog communities dominated by grasses and sedges. Because numerous species of pitcher plants, genus *Sarracenia*, are found in such bogs along the Gulf of Mexico, they are thought to be the ancestral home of *S. purpurea*, the only species found in the north.

The bald cypress, *Taxodium distichum*, hovers over wooded swamps and bogs in the South. Its spreading limbs, draped with the ghostly fingers of Spanish moss, seem to loom above the black, tannin-filled water. Roots as big as fire hoses penetrate deep into the mud and send up buttresses, called "knees," to surround the swollen base of the main trunk like a palace guard. The purpose of these "knees" is not precisely known, for the old idea that they supplied oxygen to submerged roots has not been supported by anatomical research. Bald cypress is a gymnosperm of the family Taxodiaceae, a stately group that also includes the California redwoods. Unlike most conifers, it is deciduous. An ancient specimen may be over a thousand years old and a hundred and fifty feet tall, although such venerable trees are rare because most of their habitats were logged out during the early years of this century. Spanish moss, *Tillandsia usneoides*, is not a true moss but a flowering plant of the bromeliad or pineapple family, Bromeliaceae. It is *epiphytic*, which means it grows upon another plant for support, but it is not a parasite.

Because the aerating effect of flowing water is absent, bog plants must subsist on oxygen levels even lower than that of other wetland plants. In addition, many bogs have a pH factor as low as 3.5, only half a point higher than vinegar. The pH factor is calculated logarithmically, so each point on the scale is ten times higher or lower than the one preceding it. Therefore, a soil with a pH of 3 is ten thou-

PITCHER PLANT
Sarracenia purpurea

sand times as acid as one registering a neutral pH of 7, the preferred range of most garden plants. The varied species of soil bacteria responsible for the decomposition of organic material cannot live in such an acidic environment, so dead bog vegetation does not decay but rather accumulates as peat. Peat bogs are so sterile that lumber from fallen trees is usable after centuries of immersion and ancient seeds removed from bogs sprout when planted. Even bodies have been retrieved undecomposed.

This talent for preservation makes bogs particularly interesting to plant historians. Each minute pollen grain that is released over the land every summer has a tough outer coat. When encased in layers of sterile peat, this shell has proven so durable that it is possible to identify the exact species that produced the pollen thousands of years ago. Thus core studies of accumulated peat at the bottom of bogs reveal when vegetation patterns have changed, indicating major shifts of climate.

Buoyant, quaking mats of absorbent sphagnum moss are a frequent characteristic of northern bogs. That these mats are also prevalent in the relict bogs of southern New England, the Pine Barrens, and the Appalachians tells us that these are northern communities that were originally forced south by the glaciers. Sphagnum is less common in the grass-sedge bogs of the South, where soil acidity is caused by mineral compounds, not accumulating peat.

Sphagnum is a true moss, or bryophyte, and thus lacks roots, true stems, and leaves. It forms a tangled, pale green cushion, sometimes up to a foot thick over the surface of the bog. The distinctive color results from special cells that lack chlorophyll but are able to absorb up to sixteen times the plant's own weight in water. There is also a type of bacteria that grows on the plants that combats infection so effectively, it was once used to dress the wounds of soldiers. This antibiotic effect, plus the moss's ability to retain water, is what makes sphagnum such a useful tool in horticulture. Seeds started in the moss stay moist but do not "damp off" from fungus disease.

Most plants of the open bog grow directly in this sphagnum cushion, adapting to the acidity and lack of soil nutrients in a variety of curious ways. The most obvious examples are the carnivores, a peculiar group of "insect-eating" plants of which the pitcher plant, *Sarracenia purpurea*, is often the most conspicuous.

Pitcher plants have thick, rubbery, tube-shaped leaves with a ruffled lip. These tubes, or "pitchers," collect rainwater. Glands on the lip secrete a nectar that attracts flies like carrion does, but a fly that ventures inside is immediately doomed. Short, bristlelike hairs around the lip ease his way downward, but effectively prevent him from ever climbing out again. The fly drowns in the rainwater at the bottom of the tube, and the nutrients released by the decomposition of his body are absorbed by the plant's tissues.

In June, the pitcher plants bloom, each flower appearing on a separate leafless stalk. The deep red-purple petals are short-lived, dropping after a day or two, but the curious umbrella-shaped style, or female organ, surrounded by a leathery calyx, remains on the plant all winter, giving it the look of an ingenious red-brown pinwheel.

Sundew, *Drosera* spp., is another insectivorous plant that is less conspicuous but more widespread than pitcher plants, occurring on acid, sandy soil throughout the Coastal Plain as well as in northern bogs. Most sundews form tiny rosettes of spoon-shaped leaves, although one Coastal Plain species, *D. filiformis*, has leaves like slender hairy caterpillars. This plant was one of several North American species included in Charles Darwin's book *Insectivorous Plants*, published in 1875. The leaves of all sundews are tipped with sticky glands that attract gnats and midges, snaring them like flypaper. The glutinous hairs then wrap themselves around the captive and digest him. Sundews are annuals and have tiny white, pink, or purple five-petaled flowers in summer.

A third carnivore, bladderwort, *Utricularia* spp., forms a mat of fine, threadlike leaves just below the surface of the water. Attached to them are minute round bladders, each the size of a pinhead, with lids that close on pond insects like tiny trap doors. The flower, which superficially resembles a snapdragon, blooms in June, rising above the water on a wiry stem less than a foot tall. The color may be yellow or pink. There are many species of bladderwort in the East; some grow in the still water of lakes and ponds, while others prefer bogs.

Perhaps the most famous carnivore of all is the Venus' flytrap, *Dionaea muscipula*, a plant that snaps at its prey like an animal. The instant an insect touches the hairlike triggers on the upper surface of the leaf, the two lobes snap together, snaring the hapless victim, whose struggles thereafter only serve to stimulate the production of digestive juices. The leaf stays closed for several days until all the soft parts of the animal are consumed. Venus' flytrap only grows in a small area within fifty miles of Wilmington, North Carolina, and its populations have been seriously threatened by unscrupulous and unlawful collecting.

Contrary to popular ideas, carnivorous plants do not live on insects in the same way carnivorous animals live on their prey. All these plants contain chlorophyll and carry on photosynthesis as their primary source of energy. The bodies of the insects merely supply nutrients missing in the acid environment of the bog and thus act in much the same way as a fertilizer.

Throughout the world the orchid family, Orchidaceae, is one of the largest families of flowering plants. Most orchids are tropical; of those that do grow in temperate climates, most are found in bogs. The flowers of orchids range from the inconspicuous to the flamboyant, but all have a bilateral arrangement of three petals with one, called the

PINK LADY'S-SLIPPER
MOCASSIN FLOWER
Cypripedium acaule

ROSE POGONIA
Pogonia ophioglossoides

"lip," differing from the others in size, shape, or color and sometimes in all three. Yet this is not always easy to sort out, for the sepals of orchids often resemble petals and may combine with the lateral ones to form a hood. A center column, fused from the style and stamens, is also typical of orchids, but this too may be brightly colored and resemble a petal. A further peculiarity of orchid flowers is their propensity to twist on their stems during development, so that the "lip" may start out as the top petal but end as the lowest.

Platanthera, a genus containing the rein and fringed orchids, contains most of the species found in the bogs of northern New England, but lady's-slippers, *Cypripedium* spp. (which are also found in moist, acid woods), rose pogonia, *Pogonia ophioglossoides*, swamp pink, *Arethusa bulbosa*, grass pink, *Calopogon tuberosus*, and ladies' tresses, *Spiranthes* spp., may occur, too, as well as in more southern bogs.

Orchids have adapted to the bog environment in quite a different way from the carnivores. A group of fungi species called *mychorrizae* live symbiotically on their roots, as they do on many acid-loving plants. These fungi act as stand-ins for soil bacteria, decomposing organic matter to make essential nutrients available to the plant. This delicate relationship is the reason that so many orchids are impossible to transplant, although thoughtless gardeners cannot seem to resist digging them up whenever they find them. Alas, when disturbed, the fungus invariably dies, and the orchid soon follows.

Cotton grass, *Eriophorum virginicum*, is a sedge with a tuft of silky hairs on the tip of a wiry stalk that dots the bog's sphagnum mat in summer. It is a plant of the Arctic Tundra that reaches the southern limit of its range in the bogs of the New Jersey Pine Barrens. Bog bean, *Menyanthes trifoliata*, is a creeping plant of cool northern bogs that spreads throughout the wet peat. It has a six-inch stem tipped with a raceme of delicate white flowers, each with a brush of glistening hairs on the upper surface of the petals.

The roots of bog bean are used in Scandinavian countries to make a starchy cake called "missen bread."

Pipewort, or hatpins, *Eriocaulon septangulare*, is another northern plant that forms sedgelike tussocks in the bogs of the Coastal Plain. The minute white flowers cluster in a tight sphere at the tip of a thin stem which indeed gives the plant the look of an old-fashioned hatpin. *E. decangulare* is a similar southern species. *Rhexia* spp., the meadow beauties, are southern plants that occur in sandy bogs throughout the Coastal Plain. They have delicate pink, four-petaled flowers with conspicuous yellow stamens. The seed pods are shaped like tiny urns.

MARYLAND MEADOW
BEAUTY
Rhexia mariana

Orange milkwort, *Polygala lutea*, looks like bright clumps of orange lollipops in the sphagnum mat of Coastal Plain bogs. The blossom is a tight head of tiny flowers, rather like a clover, and extremely long-lasting, for it continues to bloom from the center as the faded petals drop from the outer rim. Milkwort is a biennial, so for a colony to persist, conditions must be right for it to reseed itself.

Extremely rare, but intriguing, are the bog bunchflowers, primitive members of the lily family, Liliaceae. Tradition says that the lilies growing in the Elysian fields, that paradise of happy souls of Greek mythology, were bunchlilies. If so, these are some New World relatives. Two species, representing two genera, grow in southern bogs: the bog hyacinth, *Helonias bullata*, and the bog asphodel, *Narthecium americanum*. Bog hyacinth has tightly clustered flowers of a dull pink, each with blue stamens, that rise on a stout stem from a basal evergreen tuft of wide, straplike leaves. Staten Island apparently represents the northernmost limit of this plant on the Coastal Plain. It also grows in mountain bogs from Pennsylvania into Georgia. The bog asphodel has yellow flowers, also bunched at the top of the stalk, and thin, grasslike leaves. It grows in an isolated community in the New Jersey Pine Barrens and is not found again until the mountain bogs of North Carolina. Such a

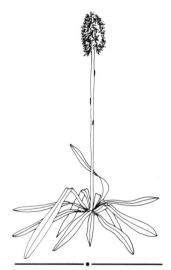

SWAMP PINK
BOG HYACINTH
Helonias bullata

widely disjunct community usually indicates a plant of ancient lineage, a survivor of another age.

The shrubs that surround a bog, sometimes forming islands in the sphagnum mat as well, are predominantly members of the heath family, Ericaceae. Heaths are boreal plants, at home in the acid podzol soil of the northern coniferous forest, where they grow on upland sites as well as in bogs. South of the boreal belt, however, such cool acid conditions exist only in bogs, where these shrubs ring the edge in remnant communities of a once colder time.

Leatherleaf, *Chamaedaphne calyculata*, is a low, spreading shrub common around bogs. It is more or less evergreen with narrow, elliptic, minutely scaly leaves with a yellowish brown cast to their undersides. The flowers are small and white with the typical bell shape of many heaths. With it grows the Labrador tea, *Ledum groenlandicum*; its specific name, meaning "of Greenland," hints of an arctic distribution. It too is low-growing, seldom reaching three feet, and has terminal clusters of white flowers in May and June. The leaves, which are smooth on top and brown and fuzzy below, are borne on twigs that are covered with tawny hairs.

Bog rosemary, *Andromeda glaucophylla*, is a choice plant of northern bogs. It forms a compact gray-green mound covered with frail pink bells in May. Bog laurel, *Kalmia polifolia*, is a rather straggly shrub with sparse purple-rose blossoms of a shape immediately recognizable to everyone familiar with mountain laurel. Its leaves are white on the undersides. Lacing the sphagnum mat beneath these shrubs are the cranberries, *Vaccinium macrocarpum* and *V. oxycoccos*. Cranberries are in the same genus as blueberries yet appear quite different. They are tiny creeping evergreens that grow right in the moss. The flowers have four recurved petals, and the stamens hug the style like the bill of a crane. In fact, the name cranberry is thought to be a corruption of craneberry. When seen in the wild, the familiar acidulous red berry looks much too large for the rest of the plant.

Cranberries are extensively cultivated on Cape Cod and in the Pine Barrens and are among the few commercial crops native to North America. Others are blueberries, black walnuts, and pecans.

Shrubs of the heath family typically have leathery leaves, often with rolled edges. They may be fuzzy on the underside, like Labrador tea, or have a waxy bloom, like bog rosemary and bog laurel. Both attributes guard against water loss, because in spite of the soggy environment, the extremely acid water of the bog is not readily absorbed by plants.

Tamarack, *Larix laricina*, and black spruce, *Picea mariana*, are trees with worldwide distributions, occurring throughout the northern coniferous forest, but like the ericaceous shrubs, they tend to congregate around bogs in the southern part of their range. Tamarack is a deciduous conifer, dropping its needles in winter and sprouting new pale green ones every spring. It is not a valuable timber tree, whereas black spruce is logged extensively for pulpwood over much of its range. A viscous black resin exuded from the bark of this tree was once known as the lumberman's chewing gum.

These tree species are replaced in the bogs of southern New England and the Coastal Plain by the Atlantic white cedar, *Chamaecyparis thyoides*. While cedars are slow growing and may live nearly a thousand years, in most places they have been extensively cut. Because the wood is both lightweight and decay-resistant, it has been used since the eighteenth century for shingles. Peter Kalm wrote in 1749 that the lumbermen of southern New Jersey were exploiting the white cedar to such an extent that he feared it would soon be completely gone from the country.

All bogs are successional communities, gradually becoming wooded swamps and, eventually, woods. In the cool northern climate this process takes hundreds of years, but in the South it would happen very quickly were it not for fire. Summer lightning fires were once common in southern bogs, sweeping across the dry surface grass in an undulating line of flame that checked the growth of woody plants but left the roots of bog plants safe in the moist peat. Where these fires have been controlled, the bogs are being invaded by thickets of wax myrtle, *Myrica cerifera*, inkberry, *Ilex glabra*, and sweet bay, *Magnolia virginiana*. The increased transpiration rate of these woody plants slowly reduces the moisture in the bog, causing the original bog plants to disappear.

·E·I·G·H·T·

Dry Lands

A beach is built by the sea. When a wave washes soil from the shore, the roiling motion of the water floats out the lighter silt and carries it seaward, but the heavy sand is dropped as the wave breaks. A sand bar begins to form, emerging at low tide as a sand spit. The waves continue to break over it, piling up more and more sand. When the sand dries in the sun, the wind takes over, swirling the grains into patterns and sweeping them along in a moving film just above the ground. As soon as this film of sand encounters a piece of driftwood or a rock, the dune is born.

DUNES

A sand dune is at the mercy of the wind that continually builds it, wears it down, and builds it again. Whatever steadiness it ever acquires is imposed on it by the dune grass, *Ammophila breviligulata*, which binds it like Gulliver with long underground stems called rhizomes. These rhizomes put down anchoring roots and send up clumps of waving grass every few feet, gradually carpeting the landward side of the dune and venturing over the crest to a point just beyond the reach of storm waves. Being buried by blowing sand seems only to stimulate the grass to grow; its

FIELD AND FOREST

SANDWORT
Arenaria peploides

SEA ROCKET
Cakile edentula

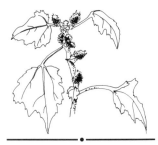

BEACH CLOTBUR
SEA-BURDOCK
Xanthium echinatum

leaves sometimes reach through as much as a yard of sand to the light. The waving clumps of grass moderate the force of the wind and insulate the surface of the dune from the blazing sun, a fact well known to children who scamper barefoot across a burning beach to the cool, grass-covered dunes.

In spite of ample rainfall and moisture-laden fog, growing conditions on a sand dune approximate those of a desert. The porous sand holds little water, and the summer combination of hot sun and wind constantly threatens plant tissues with desiccation. Consequently many beach plants look like desert plants. They have succulent stems and leaves with surfaces as hard and shiny as plastic—evolutionary stategies that conserve water and slow evaporation.

For instance, the foliage of seaside goldenrod, *Solidago sempervirens*, is noticeably smoother and thicker than its landlocked relations. Sandwort, *Arenaria peploides*, a beach-loving member of the pink family, Caryophyllaceae, forms fat clumps of rubbery leaves and stems on Atlantic beaches, producing tiny flowers in the leaf axils in early summer. Sea rocket, *Cakile edentula*, is a succulent member of the mustard family, Cruciferae. It is an annual with inconspicuous purply pink flowers in late summer and seed pods that are shaped like tiny rockets. When the seeds are ripe, they roll into the water, sometimes floating for months before washing up on another beach.

Dune plants are often prickly, too. Beach clotbur, *Xanthium echinatum*, has seed pods as spiny as a hedgehog. So does the jimson weed, *Datura stramonium*, an alien weed of dry pastures that also frequents beaches. Prickliest of all is the salt spray rose, *Rosa rugosa*, a plant that forms large colonies on the landward side of dunes in New England, where its thorny stems, wrinkled matte foliage, and blowsy flowers are a familiar sight in summer. The flowers, which come in all shades of pink as well as white, are followed by bright orange hips that look like cherry tomatoes. *Rosa ru-*

gosa is a native of Asia. It arrived, so the story goes, after a ship carrying seeds and plants for an orchardist in Boston was wrecked off Cape Cod in the nineteenth century.

Halophytes, or salt seekers, of the beach include two members of the goosefoot family, Chenopodiaceae: sea blite, *Suaeda maritima*, and orache, *Atriplex patula*. The fact that these two rather unprepossessing plants have common names at all probably means that they were once gathered for food. Most members of the goosefoot family are edible: more familiar examples are beets, spinach, and swiss chard. Saltwort, *Salsola kali*, is an annual halophyte whose seeds apparently float in on the tide, for the plant appears in late summer as a dense stand just above the wrack. It is a stiff plant that is defensively prickly: each pale, waxy-green, succulent leaf ends in a sharp thorn.

SALT SPRAY ROSE
Rosa rugosa

Other plants have other strategies. Both beach pea, *Lathyrus maritimus*, and dune grass fold their foliage on dry, windy days. Beach pea clasps its leaves along the stem like so many praying hands, deflecting the rays of the burning sun, while those of dune grass fold along the midrib, closing pleatlike ridges on the inner surface against the loss of moisture. Such plants as beach wormwood, *Artemisia stelleriana*, called dusty miller in the North, and *Centaurea cineraria*, called dusty miller in the South, are covered with minute white hairs that both reflect the hot sunlight and provide insulation for the leaves. Both these plants are naturalized aliens, the first from northeast China, and the second from the Mediterranean.

BEACH PEA
Lathyrus maritimus

The gray-green foliage of horn poppy, *Glaucium flavum*, superficially resembles that of dusty miller but sends up flowering stalks that are crowned by large yellow poppies in summer. The plant's common name comes from the long sickle-shaped seed pods that follow. Horn poppy is also an alien and seems to prefer stony beaches to sandy ones.

Beach heather, *Hudsonia tomentosa*, forms gray-blue

BEACH WORMWOOD
DUSTY MILLER
Artemisia stelleriana

HORN POPPY
Glaucium flavum

mats like juniper with needlelike leaves that cling to the stem like scales. No juniper, however, has beach heather's tiny yellow starlike flowers that bloom from May to July and proclaim it a member of the rock rose family, Cistaceae.

In their quest for moisture, most dune plants have wide-ranging root systems. They may be taproots, spreading rhizomes, or deep-penetrating tangles like fine wire. The tops are often prostrate. Plants, such as seabeach knotweed, *Polygonum glaucum*, and the dainty seaside spurge, *Euphorbia polygonifolia*, hug the surface of the sand as if cringing from the wind. The dunes of Cape Cod, Martha's Vineyard, and Nantucket are carpeted with a low-growing plant with white woolly stems and leaves like curved stilettos. This is the sickle-leaved golden aster, *Pityopsis falcata*. In August, it is covered with small yellow daisies that dot the sand like drops of summer sunlight. It can easily be grown from seed and does well in gardens with sandy well-drained soil.

Behind the dunes, where the soil has acquired some humus, nestle hardy seaside shrubs such as bayberry, *Myrica pensylvanica*, and beach plum, *Prunus maritima*. Bayberry is easily recognized by the clusters of metallic gray berries along the stems of the female plants in late summer and fall. The pungent aroma of its leaves is immediately reminiscent of old houses and Christmas, for bayberry candles can be made from the "bloom" or wax coating on the surface of the berries. In the early days, the berries were boiled in a large vat and the wax floated off the top; it took as much as a bushel of berries to make a single candle.

Beach plum has two seasons of glory. In early spring, before the leaves unfold, it is a cloud of delicate white blossoms. By fall, small edible plums hang like purple pendants all along the stem. Sometimes beach plum grows right out of the dune in what appears to be a foot-high thicket, but is actually the top of a single bush buried in the sand. Farther inland, it grows as a small tree.

Bearberry, *Arctostaphylos uva-ursi*, is a plant of the far North that occurs along the Atlantic coast into New Jersey. It is common along Cape Cod roadsides and in hollows behind the dunes, where it forms large patches beneath pines pruned into bonsai silhouettes by winter gales. Bearberry has the pendulous bell flowers typical of the heaths, Ericaceae, followed by bright red fruits. The native Americans called it "kinnikinnik" and reportedly smoked it like tobacco. The birds also like the berries, and bearberry makes a wonderful groundcover in gardens that have the requisite sandy, acid soil.

On undisturbed coastlines a tundralike heath may develop on high ground behind the dunes. There, various species of blueberry and huckleberry, *Vaccinium* and *Gaylussacia*, mingle with pasture rose, *Rosa carolina*, shining sumac, *Rhus copallina*, and arrowwood, *Viburnum dentata* or *V. recognitum*. In August, these shrubs are interspersed with clumps of wild indigo, *Baptisia tinctoria*, a bushy plant with yellow pealike flowers. Later come the goldenrods and asters. One especially attractive species is the low-growing stiff aster, *A. linariifolius*. Sometimes one can find the delicate lavender-pink spires of racemed milkwort, *Polygala polygama*, in such places as well.

BARRENS

The word *barrens* was bestowed by the early settlers on any area that was useless for farming because of thin, nutrient-poor soil. Botanically speaking, however, these sites are far from barren, for they support some of the most interesting plant communities in the East.

∴ FIELD AND FOREST ∴

PINE BARRENS

We have already described the deciduous forest as the climax vegetation pattern for most of the eastern United States. Yet there is a second pattern that some botanists call an alternate climax: the pine, pine and heath, or mixed pine and oak forests that occur on poor sandy soils. Such so-called pine barrens are usually associated with the Coastal Plain, but they may also occur in the Piedmont on steep, south-facing slopes and along sandy river bluffs, and even on thin, gravelly soil in the Appalachian mountains, the evolutionary parent of all eastern plant communities. In spite of adequate rainfall and a generally humid climate, all these locations are subject to punishing drought because of the inability of sand to retain moisture. Excessive dryness also makes fire a constant presence. In fact, this forest burns on a regular basis. If it did not, deciduous trees eventually would take over, so perhaps it is not truly a climax community, but a successional stage that has been halted by fire.

Pine barrens exist all along the Atlantic and Gulf Coasts of the eastern United States, but the various species of pine differ according to the latitude. The northern pitch pine, *Pinus rigida*, is the dominant species in pine barrens of Cape Cod, Long Island, and New Jersey. New Jersey also represents the northernmost distribution of loblolly pine, *P. taeda*, and shortleaf pine, *P. echinata*, both trees that are common in barrens to the south. Longleaf pine, *P. palustris*, is found south of Virginia and is the dominant pine of the Sandhills area of North Carolina. Slash pine, *P. elliottii*, is the most southern species, occurring from South Carolina to Florida and the Bahamas.

Of all these locations, the New Jersey Pine Barrens are the most botanically diverse. The combination of lowland bog plants, found along the numerous slow-moving streams, and the upland plants of dry areas, plus the combination of northern and southern species, make it unique.

So unique, in fact, that up to 12 percent of the plants that grow in the Pine Barrens are considered rare or endangered. In addition, these woods have been subjected to every human pressure typical of most pine barrens in the East.

In December of 1632, the navigator of a Dutch boat standing off the coast of Cape May wrote in his diary that the land gave off a "sweet perfume . . . [that] comes from the indians setting fire . . . to the woods and thickets, in order to hunt." This is the earliest reference to fire in the New Jersey Pine Barrens, an incongruous sandy wilderness of pine and oak occupying nearly one and a half million acres of the outer Coastal Plain. Only a few miles to the west of these woods is a network of interstate highways carrying an unbroken line of traffic north to New York City and south to the Delaware Memorial Bridge. Philadelphia is less than an hour's drive to the west, and to the east is Atlantic City and its satellite resorts on the Jersey shore. Yet within this forest the population is somewhere around fifteen people per square mile.

People exploit the barrens, yet for the most part, they do not live there. Even those tribes who burned the woods in 1632 probably came from outside villages, although their ancestors had been hunting in the pines since the Pleistocene Ice Age. The Europeans, in their turn, pronounced these woods only fit for cutting wood and running cattle. Farms were small. Yet plots were not shifted around as they were elsewhere, for it took hard work and quantities of manure to make the sandy soil produce at all. Most farmers didn't bother, but simply burned off the surrounding woods and turned out their cattle and horses to graze. No one protested, for in the early days, most of the land was still owned by the West Jersey Society in England. In the 1820s and 1830s, farming in a more traditional sense did experience a small boom when it was discovered that glauconite, also known as "greensand marl," a sediment plentiful on the inner coastal plain, made an excellent fertilizer for sandy soils. But like most Pine Barrens enterprises, it didn't last.

The few permanent citizens of the pines have always been forced to pick up a living where they could, and one way or another, this has usually meant cutting and burning the woods. Blueberries grow wild in the pines. In the days before they were cultivated and the practice forbidden, the bushes were regularly set afire because those that came back afterward always bore the best crops.

Bog iron occurs when acidic water precipitates soluble iron salts from porous soil, and the smelting and working of bog iron was a thriving industry in the pines from the years before the Revolution into the early nineteenth century. Sparks from the forges and furnaces often set the woods on fire, and sometimes they were burned deliberately. Owners of timberland rarely lived in the barrens, and charred trees that were useless for lumber could always be bought cheaply for the charcoal that was needed to fuel the iron industry.

The bog iron was largely depleted by the midnineteenth century, but by 1900 sparks from railway engines were igniting fires in the woods. The railroad brought the pines within reach of New York, causing the real estate developers to move in. They bought thousands of acres from the defunct iron companies to sell off in small lots. Yet few buyers actually moved to the barrens, and those that did found the summers hot and humid and the mosquitoes unbearable. They didn't stay, and the promoters eventually went broke and sold the land for back taxes.

Today clearings are scattered among the pines where the limestone foundations of vanished buildings have altered the pH of the soil sufficiently to grow grass and poison ivy. Catalpa trees planted long ago still surround a vanished town called Martha, along with a disjunct community of a rare green orchid called Loesels twayblade, *Liparis loeselii*.

∴ DRY LANDS ∴

In 1978 most of the land comprising the Pine Barrens was put in the New Jersey Pinelands Natural Reserve, subjecting all future development to a limited orderly plan. Yet the fires continue to burn: set by the casual cigarette, the undoused camp fire, or arson. All told, there are over a thousand fires every year, and only about ten of them are caused by lightning. Most places average one fire every ten to thirty years, but in an area called the Plains, the woods have been burned so often that the pines and blackjack oaks have formed peculiar multistemmed dwarfs, none taller than a man, that grow dense tops from a massive root crown called a stool.

The oaks that grow in frequently burned areas are fire selected: Post oak, *Quercus stellata*, blackjack oak, *Q. marilandica*, and bear oak, *Q. ilicifolia*. All have leaves so full of protective oil they are as thick as polished leather. Where the woods have escaped the flames for fifty years or more, the pines have been invaded by black, white, chestnut, and scarlet oak, with a lesser proportion of red maple, gum, and hickory.

Only three species of pine are able to sprout from the trunk after a fire, and two of them, the pitch pine, *Pinus rigida*, and shortleaf pine, *P. echinata*, are widespread in the Pine Barrens. The third is a native of Mexico. Pitch pine is a northern tree, growing from coastal Maine down the spine of the Appalachians, where it is always found on the thin, gravelly soil of steep slopes, ridges, or plateaus. Shortleaf pine is southern, a native of the Atlantic Coastal Plain. They are but two examples of numerous northern and southern taxa that meet in the barrens. To understand how this happened, it is necessary to go back several million years.

The oldest soils of the Pine Barrens were laid down sometime in the Cretaceous period, about one hundred and forty million years ago, when dinosaurs walked the earth and flowering plants had just begun to mingle with coniferous species already a hundred million years old. Ancient pollen deposits from other locations on the Coastal Plain show that early oaks, heaths, bayberries, and members of the magnolia family already existed in the Cretaceous era, in fact, all the principal species that make up the dominant vegetation of today's Pine Barrens.

Yet the outer Coastal Plain, where the barrens are today, was to disappear and reappear ten times or more. Ancient seas washed over the land and then receded, boiling up the soil and laying down new layers of sediment to be cut by flowing water and roiled by ancient winds. About five million years ago, the land stabilized, and pollen evidence shows that by a million years ago, the climate was so warm that some trees and shrubs that grow in the barrens today were mixed with southern or even tropical species. The northern pitch pine was not present at that time. It moved southward ahead of the glaciers, which were still to come.

The sheet of ice known as the Wisconsin Glacier crept to within forty miles of the northern edge of today's Pine Barrens, forcing tundra species such as broom crowberry, *Corema conradii*, to move south. According to pollen layers from that time, both the pitch pine and other pines of the boreal forest like the jack pine, *Pinus banksiana*, and red pine, *P. resinosa*, were present in the barrens. After the glaciers receded and the climate warmed once again, the oaks and southern pines reinvaded. Positioned as it was on the leading edge of shifting climates, the Pine Barrens thus acquired its unique mixture of northern and southern plants.

Because vast quantities of the earth's surface water froze into the glaciers, much of the continental shelf was exposed during that time. Botanists theorize that plants may have migrated north across this land, their seeds carried by swirling sou'westers born of subtropical air masses. Many species that colonized the barrens during this period apparently became isolated, for some species there today grow nowhere else on earth. Others are disjunct populations, meaning they are not found again for hundreds of miles.

The soil of the Pine Barrens is best described as dirty sand. It is a northern podzol soil, out of place in New Jersey, a soil so sandy and porous that it swallows water immediately, leaching out the nutrients and leaving the surface as dry as a dune. Everywhere, the woods floor is covered with a litter of undecomposed oak leaves, pine needles, and cones, because with a pH of 4, the soil is far too acid for earthworms or bacteria. Between fires, three to four inches of this litter may accumulate on the forest floor, while the laurel, blueberries, and other shrubs grow into a tangled, resinous, inflammable mass several feet high. In summer, droughts are common, and a spark landing on dry litter will cause the waxy leaves of laurel and heaths to flare like tinder, igniting the pines above them into crackling torches.

Thus the fire and the soil have selected the vegetation of the Pine Barrens, favoring plants that are both highly inflammable and yet able to recover quickly after being burned. The pitch and shortleaf pines have thick bark that protects them from extreme heat, and within a surprisingly short time, dormant buds will sprout from blackened trunks and limbs and new green shoots will rise from the roots. The three typical oaks of frequently burned areas, the post, the bear, and the blackjack, are all fire resistant to varying degrees, yet only the blackjack oak, *Q. marilandica*, seems to actually thrive on fire and will come back even after the oldest pines have been killed. Then for a time it becomes the predominant tree of the woods, while all around it, germinating pine seeds sprout from the charred earth.

Two races of pitch pine have evolved over the millennia. The trees of the first race produce cones that open when they are ripe in the normal manner of most conifers,

BEARBERRY
Arctostaphylos uva-ursi

TRAILING ARBUTUS
Epigaea repens

but the cones produced by the second race remain closed, sometimes for years, until they are exposed to the searing heat of fire. Both kinds occur throughout the Pine Barrens, yet all the pines in the dwarf forest of the Plains belong to the second race.

Like the plants of the dunes, many plants that grow on sandy, acid soil tend to form tufts or mats. Bearberry, *Arctostaphylos uva-ursi*, that plant of dune and tundra, also grows here. The diminutive pyxie moss, *Pyxidanthera barbulata*, is a tiny alpine plant from the Appalachians that forms tufted cushions on the sand, sending out radial branches that root in the undecomposed litter. In April, pyxie moss is covered with round pink or white flowers, each with a fused corolla that splits into five triangular petals. In the pines, it often grows with trailing arbutus, *Epigaea repens*, another creeping evergreen, and sweet fern, *Comptonia peregrina*. Sweet fern is not really a fern, but a shrub that forms dense thickets three to four feet high in burned-over areas. The Victorians grew it in their gardens for the pungent fragrance of its leaves. Bracken, *Pteridium aquilinum*, is a true fern of worldwide distribution with a root system so deep it is unaffected by fire. It too is common in the pines.

Sand myrtle, *Leiophyllum buxifolium*, is another mat-former that is found in the Pine Barrens: a diminutive tuft of upright branches with minute leaves and pink or white flowers in loose terminal clusters. It often grows with *Hudsonia ericoides*, a close relation of the beach heather, *H. tomentosa*. This species is greener. It looks less like juniper and more like a club moss, except for its tiny yellow flowers.

The most widespread and conspicuous shrubs of the Pine Barrens, in both wet and dry habitats, are the heaths, family Ericaceae. In dry, sandy areas, black huckleberry, *Gaylussacia baccata*, and low-bush blueberry, *Vaccinium vacillans*, form an almost continuous knee-high blanket beneath the pines and the oaks. The dangleberry, *Gaylussacia frondosa*, is also common. To tell a blueberry from a huckleberry, look for small resin glands, like amber dots, on the underside of the leaves. If present, the plant is a huckleberry.

The laurels also grow everywhere, both mountain laurel, *Kalmia latifolia*, and the smaller, purple-pink sheep laurel, *K. angustifolia*, a plant that was once destroyed over much of its range because it is poisonous to cattle and sheep.

Turkeybeard, *Xerophyllum asphodeloides*, is a distinctive plant of the Pine Barrens and a member of the legendary bunchlily family. It blooms in June on soil that is dry on the surface but moist beneath, a condition that is not uncommon here, where aquifers frequently approach the surface. Turkeybeard is a stately plant with spires of white flowers on sturdy three-foot stems rising from a hummock of grasslike leaves. It is a southern plant that originally migrated from the Appalachians onto the Coastal Plain and reaches its northernmost limit in the Pine Barrens.

Pink catchfly, *Silene caroliniana*, forms a tidy clump of flowers every spring. It is a member of the pink family, Caryophyllaceae, and pretty enough to grow in any garden. *Chrysopsis mariana*, the Maryland golden aster, is a handsome yellow composite with yellow rays and yellow disks that blooms on a stout, silky stem about two feet tall. It flowers in the Pine Barrens during August and September, and while it looks like a yellow aster, it is botanically closer to the goldenrods. The genus name translates as "golden face."

In the fall the silvery aster, *Aster concolor*, a species with conspicuous fuzz on its leaves and stem, blooms here. It is listed as an endangered plant. More common but no less beautiful is *Aster spectabilis*, a plant that grows in the barrens with bearberry and sand myrtle. Last of all is the lovely Pine Barrens gentian, *Gentiana autumnalis*, also called one-

TURKEYBEARD
Xerophyllum asphodeloides

flowered gentian. It is a slender plant, about a foot and a half tall with very narrow, delicate leaves. The color is usually the typical deep blue of the gentians, but sometimes an isolated plant may be white or purple.

SERPENTINE BARRENS

Even before the land was settled there were clearings in the forest like miniature grassy prairies. Some of these ran from Philadelphia to the Susquehanna River in an irregular line that when viewed from above must have broken the sea of trees as a serpent's coils might smooth the surface of the water. These misplaced bits of prairie were called the serpentine barrens, named not for my metaphor, but for the underlying, peculiar green rock. This same rock occurs in northern Italy, where it harbors a small snake of the same color.

Serpentine barrens occur in many parts of the world. Besides Italy, they are found in Cuba, Japan, and Norway. Some geologists think they may mark boundaries where tectonic plates ground together and forced molten serpentine through the surrounding bedrock.

Because the soil is thin over such rock outcroppings, it dries quickly, so dryness is one factor in selecting plants for these sites. Yet it is not the only one, for the soil over serpentine lacks phosphorus, nitrogen, and calcium, all nutrients necessary for the health of most plants. The aging rocks also release chromium, nickel, and magnesium as they decompose. These metals are toxic to some plants and may have a dwarfing effect on others. Such a highly specialized habitat selects specialized plants, so it is not surprising that these barrens support some of our rarest American natives.

The most familiar rare plant of the barrens is moss phlox, *Phlox subulata*, which in early spring paints hundreds of eastern lawns and garden walls a vivid shade of hot pink. In the wild, the pink seems far less strident, yet it is the

Moss PHLOX
Phlox subulata

reason that locally many serpentine outcrops are called "pink hill." The phlox often coats the steepest part, growing in a sort of green scree with another serpentine endemic, *Cerastium arvense* var. *villosum*, a particularly handsome species of the lowly chickweed. The word *villosum* means "covered with soft hairs" and alludes to the fuzzy gray-green foliage. The white flowers are the size of a dime and have the notched petals typical of chickweeds.

The slender tufts of rock sandwort, *Arenaria stricta*, grow on these barrens, and a relative of portulaca called sunbright, or fame flower, *Talinum teretifolium*, a plant with wiry stems and pink flowers that last less than one day. In summer the serpentine aster, *Aster depauperatus*, blooms, a bushy species that is covered with hundreds of tiny white flowers. Whorled milkweed, *Asclepias verticillata*, has long, slender leaves with the texture of rosemary. They are arranged in whorls around the stem, and in summer creamy white flowers bloom in the axils. Serpentine species of familiar genera often seem dwarfed, and this is true of the dry land snakeroot, *Eupatorium aromaticum*, a smaller, more delicate version of the familiar snakeroot of our autumn woods.

In recent years development has destroyed many serpentine barrens, and pollution and acid rain have altered the soil of others, allowing multiflora rose and black locust trees to crowd out the indigenous flora. Burning apparently controls this to some degree.

LIMESTONE BARRENS

Outcrops of limestone, a gray and white sedimentary rock composed mainly of calcite, produce soils of a neutral pH unusual in the eastern woods. While neutral soils are extremely fertile, on these rocks the layer is too thin to retain moisture and so must be considered as a xeric, or dry, environment. Hoary puccoon or Indian paint, *Lithospermum canescens*, is a plant of the midwestern prairie that sometimes

HOARY PUCCOON
Lithospermum canescens

occurs on limestone barrens in the East. It forms a showy clump of oblong leaves that are covered with hairy fuzz and support brilliant orange-yellow flowers. The word "puccoon" was given by the Indians to many plants that were used for dye. Cliff green, *Pachistima canbyii*, also occurs on limestone barrens. It is a low-growing evergreen shrub much valued by gardeners as a groundcover.

SHALE BARRENS

Some Appalachian ridges have slopes that are thickly strewn with slabs and flakes of gray-brown shale. Geologically these outcrops date to the middle Devonian period, the age of the fishes, and date from about three hundred and twenty million years ago. They occur most often in Virginia and West Virginia, but there are some examples in western Pennsylvania as well. Shale barrens support some of the rarest plants of the East, including Kate's mountain clover, *Trifolium virginicum*, and shale morning-glory, *Convolvulus purshianus*. *Aster oblongifolius* is a rather stiff, bushy plant from the Midwest that sometimes blooms on shale barrens. Its flowers are lavender and among the more handsome of the aster tribe.

III
Ecological Landscaping

·N·I·N·E·

Your Garden as a Plant Community

So many patterns, so many diverse plant communities enliven the American landscape, why is so little of this richness reflected in our gardens? Our eastern woods and fields once had great diversity; a diversity that we are losing bit by bit, each time a woods is cut, a wetland drained, or a field replaced by pavement or lawn. Now there is concern in some areas that even those communities that have escaped overt destruction may no longer be able to reproduce themselves because of competition from alien pest plants and the long-term effects of altered drainage patterns.

At present, our gardens and public plantings do almost nothing to atone for this loss. Rather, the sights of most American gardeners and growers remain firmly focused on garden hybrids and exotic species; and that, unfortunately, is a large part of the problem. It is well known to botanists and ecologists, for example, that invading Norway maples, *Acer platanoides*, now pose a serious threat to eastern forests. Yet the Norway maple continues to be one of the most widely planted trees in North America. It is sold at every garden center and nursery and even distributed free by tree-planting organizations to combat global warming. All this despite the fact that it is not a good garden tree: Norway maples have dense foliage and greedy feeder roots that kill all but the toughest grass or groundcover. Its only vir-

FLOWERING DOGWOOD
AND BIRD
Cornus florida

tues seem to be that it is stress resistant and easy for nurseries to propagate.

The Norway maple is but one example of the lack of thought given to the danger some imported plants pose to natural landscapes. While it is true that those that have become the most invasive represent but a small fraction of the total number of exotics grown, the fact remains that several of the worst offenders were originally introduced as garden plants. The damage they have inflicted is now extremely serious: instead of the airy layers of a healthy forest, fields of waving grass and flowers, and wetlands filled with cattails, iris, and arrowhead, we have trees strangled and broken by encircling vines, fields overrun by dense thickets of multiflora rose, and wetlands completely given over to purple loosestrife or phragmites.

At the same time, it is difficult to tell from our gardens and public spaces in what part of the country we live. Over and over again, the same kinds of trees and shrubs are planted; usually fast-growing exotics with shallow root systems that are easy for nurseries to propagate. The inevitable result is that every shopping mall and housing development soon looks like every other. Apparently without noticing, we have traded a rich inheritance of diversified plant communities for the brass coinage of mass-produced shrubs and trees.

Even sadder, many of us seem to plant our yards principally to benefit the neighbors, mowing acres of lawn that we never use and restricting the use of woody plants to foundation plantings that can only be enjoyed from the street.

Why do we subject ourselves to the hours of repetitive mowing and clipping these sterile landscapes require, when they do nothing to help our beleaguered plant communities or the wildlife that depends on them? What is the point of keeping backyard bird feeders if we do not also provide the conditions birds need to survive and reproduce? How can

∴ YOUR GARDEN AS A PLANT COMMUNITY ∴

we expect to enjoy clouds of butterflies around our flowers in summer, if we do not also learn to accommodate their eggs and larvae?

As more and more of our woods and fields are replaced by lawns and pavement, it's time we gardeners asked ourselves these questions. Birds need insects, leaf litter, and fruiting shrubs, as well as dead wood, both fallen and standing, and herbaceous plants that are allowed to go to seed. Butterflies need both sweet-scented flowers and an abundance of larval food plants, and all of us need to learn how to provide such conditions in our gardens.

Until very recently, virtually all of America's gardens, great and small, public or private, were collections of commercial hybrids or selected species from widely differing locations and habitats. Now, many gardeners are coming back home, so to speak, to their own botanical heritage. In these days of environmental crisis, these gardeners feel a special responsibility, not only to protect and restore whatever native plant communities remain, but to teach the rest of us how to celebrate this country's unique natural beauty and diversity. By helping native plants to thrive in their own gardens and neighborhoods, not only are these gardeners helping to stem the loss of biodiversity, but they are bringing us all into closer contact with the natural world.

These gardeners do not battle every kind of invading organism, but allow room in their garden for other creatures besides themselves. This does not mean, however, that they live among weedy lots and ragged lawns. Their idea is not simply to allow what will happen, to happen, but to create the same kind of casual harmony that is found in unspoiled natural areas.

In a city or suburbs, you will, of course, have to consciously design such a garden. In more rural surroundings, however, you may be able to "release" it from an already existing landscape by simply eliminating certain plants, encouraging others, and adding a few as you go along.

TIGER SWALLOWTAIL
BUTTERFLY
Papilo glaucus

∴ FIELD AND FOREST ∴

There are many opportunities for such released landscapes. In parks and nature centers all over the East, armies of volunteers are already helping to restore and manage degraded woods and fields. Some highway departments, chastened by the destructive results of former policies, are managing roadsides in a more enlightened manner, and a few business executives are even installing natural landscapes on their corporate campuses. In many areas, too, clustered developments are now encouraged. The idea is to slow the loss of open space by building compact "villages" surrounded by protected open land. What better place to provide a natural landscape for growing children to explore?

Nowadays, any open land that is not actively farmed must be managed in some way, lest it become a useless tangle of Oriental bittersweet, multiflora rose, and honeysuckle. At present, most of it is unrelentingly mowed, sometimes resulting in sterile "lawns" of thirty acres or more. Quite apart from the needless destruction of precious wildlife habitat, mowing on such a large scale is simply a waste of energy, money, and time.

There is a better way. For it to succeed, however, most of us will need to know a great deal more than we presently do about the ecology of native plant communities, the soils they depend on, and the wildlife they support. We will also need to study the dynamics of succession, how best to control or promote it, as well as the safest and most effective ways to reduce competition from invasive aliens. If we are to stem the alarming loss of biodiversity, we must all do what we can to encourage such projects in our communities. As gardeners, we can also help by growing our beautiful native trees, shrubs, and herbaceous plants at home.

· T · E · N ·

Nature by Design

The first Europeans to come to this country were overwhelmed by the abundance of the wildlife, the vastness of the forest, and the space. Peter Kalm wrote in 1750 that "wherever I looked to the ground, I everywhere found such plants as I had never seen before." It is little wonder that he and others were soon sending a rich flow of North American flora back to the gardens of Europe.

No one in those days, of course, thought of plants as members of an ecological community. The idea would have been completely alien, even to the great eighteenth-century collectors like John Mitchell, John Clayton, or the Bartrams. As fascinated as these men were with the endless array of new and interesting species, they collected each plant as a discrete individual, unrelated to its companions.

Just as their counterparts do today, sophisticated eighteenth-century gardeners like George Washington and Thomas Jefferson modeled their gardens on English or European designs and filled them with a mix of European, Oriental, and American plants, each one individually selected for a particular ornamental characteristic. The original habitat or associations of these plants in the wild had little or nothing to do with the way they were combined in the garden.

Only now, in the final years of the twentieth century, has there been any thought of developing what may turn out to be a distinctly American style of gardening: one that

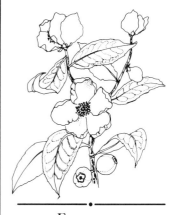

FRANKLINIA
Franklinia alatamaha

is based on indigenous plant communities. Interest in gardening is stronger than it has ever been in this country, yet many gardeners find themselves dissatisfied with the formal beds and geometric styles of traditional European garden design, perhaps because they fail to reflect the informality and enthusiasm that is so much a part of the American character.

American gardeners have also become wary of using too many garden chemicals, correctly perceiving many of them as environmentally dangerous and of little benefit in the long run. Others question the wisdom of using valuable water and energy to maintain vast acres of largely useless lawn, and reject the sort of rigid and uncompromising control that has all but destroyed the character of the wild landscape.

Despite our increasingly urban life-styles, a love of wildness still seems to occupy a large piece of the American psyche. We long to live among nature's intrinsic patterns and colors, not only for their inherent beauty and harmony, but also because, in these days of continual bad news about the environment, the idea of restoring even one small corner of a damaged landscape is profoundly satisfying. At the same time, we all want gardens that are practical, livable, and need relatively little maintenance.

However sound these intentions, a lot of us who love to "botanize" in the field are completely daunted by the very idea of designing a garden. Books on the subject are of little use: the examples never seem to suit our house or our lot. And, as much as we love native American plants, deciding which ones to choose for a garden can become so complicated that many of us simply give up and settle for whatever is available at the local garden center.

It isn't easy to make a garden, but it can be a lot of fun. You must start with the idea that your site is unique and that your job, as designer, is to uncover its inherent possibilities. At the same time, your garden should be exactly that: a private space that fits the needs of you and your family. If you want a secluded refuge, you can have it; if you want space to entertain friends, you can have that too. Nor does a garden based on regional flora mean you can't also put aside space for flowers, herbs, or vegetables if you want them.

There is, of course, more than one way to celebrate North American plants. For many city gardeners, the best course may be to simply incorporate choice natives into a more traditional garden plan; a sort of affirmative action project. Or they can try to capture the essence of a natural habitat in a spare, symbolic way, as the Japanese do. Those with a little more room may reproduce a slice of an authentic natural landscape in the garden. Still others will not be interested in gardening as much as rescuing an existing community from invasive aliens. There is room for all these approaches in the new American garden.

Whatever your ultimate intention, begin with the plants that are already growing on your lot. Decide which trees and shrubs you want to keep, what should be taken out, what might be moved. Don't be hesitant about this; all true gardens are in a continual state of renewal. If your house is surrounded by overgrown evergreens that are blocking out the air and sunlight, by all means cut them down! On the other hand, with the exception of invasive species, there is no reason why an attractive, well-mannered exotic that is already doing well cannot be incorporated into the new design.

If you don't already have an outdoor sitting area, decide on the best place for one. Go to that spot, sit in a chair, and consider the view. Is there something in the way? Do you feel the need of more privacy? Perhaps your neighbor has a tree or a vista you can take advantage of. Such "borrowed" views can be an important part of garden design.

Walk around the house. What is the best place for a children's play area, a handy place for the garbage cans? Do you need to screen out the cars? All these practical necessities should be included in your plan. Remember also to go inside and look out the windows. Are the views as interesting as they might be? What will they look like in winter?

Examine the lay of the land. Where are the slopes, the high points, the low points? From what direction are the prevailing summer and winter winds? Plot the path of sunlight in all seasons of the year.

Get your soil tested. Is it predominantly clay or sand, acid or alkaline? A healthy soil is the first requirement for a healthy garden. Remember that good soil has its own atmosphere, its own structure, and its own organisms, and should always be treated gently and with respect.

Unfortunately, it rarely is. The sort of compacted, homogenized dirt that is so often left behind in new housing developments may need substantial additions of good topsoil and humus before it can support a healthy garden. If you live on what was once agricultural land, your soil may also lack essential nutrients, particularly if it is sandy and well-drained, and may even contain pesticide residues. Soil in urban gardens, on the other hand, can be seriously compacted, as well as low in organic material, and contain building rubble to boot. For this reason, city gardeners may find it more satisfactory to construct raised beds in the garden and bring in the soil to fill them. Some urban soils, however, resemble the oxygen-poor earth found on floodplains, which explains why floodplain species (such as sweet gums, elms, and sycamores) often do well as street trees. Seaside species can also be more tolerant of salted city streets than inland species. The point is, take the time to analyze your particular situation, and try to choose plants that are ecologically appropriate.

Look carefully at microclimates as well; those areas of the garden that have the most shade in summer, for instance, or the most sun in winter. Where does the soil tend to be thin and dry because of underlying rock, or slow to drain after a rain? These conditions will also help indicate your choice of plants. Be imaginative. One coastal New England gardener grows a dune plant called sickle-leaved golden aster, *Pityopsis falcata*, in pure sand between the stones on her terrace. (She also has a luxuriant bed of moisture-loving mint growing in the outdoor shower!)

It is a good idea to keep the highest-maintenance areas close to the house (the flower bed near the sitting area, for instance, or the herbs and vegetables by the kitchen door) and increase the degree of wildness as you move outward through the garden.

Think of the lawn as an outdoor living space, and plan only as much as you will need. In most cases that will be far less than you think and, in a small city garden, may be none at all. You may find that replacing the front lawn with shrubs and trees underplanted with groundcover will provide far more privacy and seasonal interest, whereas much of the back lawn might be kept as a meadow and mown only once or twice a year. If you are concerned about the neighbors, mow paths through the tall grass; they will immediately make it look groomed rather than neglected.

For privacy, it is rarely necessary to plant all along the boundary of your lot, thus making it seem smaller and narrower than it is. Try, instead, to place groups of plants in such a way that they block certain views but allow others. At this stage you need consider only the shapes and sizes of the plants you want; a tall narrow tree here, for example, a more rounded one there, or a screen so many feet high.

When you have decided on the main elements of your garden, mark the different areas on the ground with stakes and string; a yellow nylon cord is a good choice for this because it is light, easy to see, and convenient to move around. Use color-coded stakes to represent trees and shrubs and group them according to relationships found in nature; the four layers of a healthy woods, for instance, or the kind of broad intermingling of shrubs, flowers, and grass that are found in a high-quality old field.

By now, you should have a pretty good idea of what kind of natural community is most appropriate for your site, an "old field," for instance, or a wood's edge you can enjoy from inside the house or, if yours is an older neighborhood with large trees, a bit of shady woodland. If you have a suitable area of poor drainage, you might even try a wet meadow or a section of wooded swamp. It is best to choose a model that is common in your area; rare communities such as barrens or bogs do not transfer easily to the home garden.

SASSAFRAS
Sassafras albidum

Once you have decided on the appropriate community, go into the field and study it. Not only is this an important step in the design process, it is also the most fun. Pick out a group of plants that seems especially attractive to you, and try to analyze the lines, forms, colors, and textures that make up the underlying pattern. Isolate those species that contribute most to its appeal, and notice their shapes and growth habits. In a successionary old field, for example, an upright red cedar often provides a strong vertical accent, whereas sumacs sprout from underground roots to form large rounded thickets. There may be groves of young sassafras trees, another root-sprouting species, or an isolated hawthorne or dogwood standing amid swirling grasses and field flowers.

Notice, too, which species clothe a hillside and which follow the floodplain of a stream, and think about the high and low spots of your site. Observe spaces and edges; how certain clumps of shrubs and trees create hidden clearings and peninsulas.

You will probably discover that the design you like is dependent on a relatively small number of species. The character of a woods, for instance, depends on the species of canopy trees (oak, pine, spruce, or maple, depending on where you live), and relatively few kinds of grasses will cover most of a field.

The moment you try to fit all this information into your garden plan, however, you will immediately run up against the difficult question of scale. How *does* one translate a fifty-acre woodland onto a two-acre suburban lot, or imitate a thirty-acre field in a city backyard?

Scale is, of course, intimately related to time, which is just another way of saying that you must allow ample room for the mature size of each tree or shrub that is to form the "bones" of your garden. In other words, if a tree species reaches sixty feet at maturity, you must allow space for a sixty-foot tree, even if the actual plant you install is only three-feet high. Obviously, most suburban "woodlands" will have room for only a few canopy trees surrounded by smaller understory trees and shrubs. Few city gardens will have room for even one; here, a large shrub pruned in the shape of a small tree may be the best choice. The idea is to pare down and simplify complex natural relationships.

All gardens change as they grow. The challenge is to accommodate the inevitable. One way to do this might be to plant a grove of young trees and underplant them with native grasses and field flowers to simulate a certain stage of old field succession; then, as the trees grow and their tops intermingle to produce heavier shade, you can gradually replace the grass with various woodland shrubs, ferns, and wildflowers.

Again, if you want to keep the old field character indefinitely, plant shrubs instead of trees. A group of native viburnums might be a good choice, or, on the thin soils of coastal New England, a single red cedar, surrounded by bayberry or blueberry, or underplanted with bearberry, is a striking combination. On the Piedmont, nothing is lovelier than a group of dogwoods standing amid tall grass. Broomsedge and purple top would be good choices because both turn a lovely color in the fall. Farther south, you could use green hawthorne, *Crataegus viridis*, noted for its outstanding fruits, or, if your soil is acid and well-drained, the lovely sourwood, *Oxydendron arboreum*. The possibilities are endless; your own area will supply the best choice for your garden. Go out into the field and observe what pleases you.

It will be obvious by now that designing a garden based on regional native landscapes requires an intimate knowledge of plants. You will need to be able to recognize wild species, both native and alien, at all stages of growth and in all times of the year. This is particularly true for those gardeners who are trying to reclaim a neglected field or young woods from the grip of invasive aliens. Reducing the competition from these plants can be a difficult task, but, even after years of neglect, the original natives often reappear.

At that point, you, as gardener, may choose to limit your role to providing protective maintenance while the field slowly returns to woods. Or you may choose to keep the area completely open, or allow a few islands of shrubs and early successionary trees to develop. All these are legitimate design choices.

Then, with the judicious use of pruning shears, you can learn to feature elements you particularly like, emphasizing certain shrubs, for instance, or drawing attention to the growth habits of trees. For example, the trunks of woodland trees always provide strong verticals as they lift their branches toward the light. Trees found in open fields, on the other hand, are more horizontal in their effect, while those found along a floodplain may lean and bend in interesting ways. Any of these can add drama to your landscape.

In both the created natural garden and the released natural landscape, the beauty and strength of the design come from structures and relationships found in native communities. These are gardens that feed the spirit as well as the eye. They are loud with the sound of bird song, of bees humming in the flowers, of wind in the trees or grass.

∴ NATURE BY DESIGN ∴

They are gardens filled with the tantalizing scents of flowers and foliage, as well as the rich aroma of a healthy soil. Most of all, you, the designer, will have the deep satisfaction of knowing that your garden, as an extension of the greater landscape around you, will be doing something, however small, to help restore the world.

·E·L·E·V·E·N·

Maintaining the Natural Garden

PHLOX WITH PRAYING MANTIS
Phlox paniculata with *Mantis religiosa*

All good gardens are a balancing act between the dynamic forces of nature and the inclinations of the gardener. Once your garden is designed and planted, your job will be to take care of it in a way that gently tips the balance toward the plants you want to grow and thrive and away from those you do not. As a natural gardener, you have already learned to work with nature in your choice and mix of plants; now you must learn to work with nature to help them thrive.

The biggest maintenance problem in any disturbed landscape, including the garden, is weeds. A weed is not just a "plant in the wrong place," but an opportunistic invader. Weeds seldom gain a foothold in unspoiled natural areas. They are plants uniquely equipped to grow on disturbed soil, which is why they are so successful in gardens, abandoned fields, and roadsides, vacant lots, second growth woods, and dumps. They are tough, fast-growing plants that produce copious amounts of seed. Many also have wide-ranging root systems capable of sending up shoots far from the parent plant, or stems that root wherever they touch the ground.

Because these ruderals, as they are called, originally evolved in habitats of regular and severe disturbance, they are able to colonize bare soil almost immediately. Therefore, the best way to keep them out of the garden is to disturb the soil as little as possible. Develop your garden plan slowly, never opening up more ground than you can control, and allowing the plants in one new area ample time to become established before beginning on another.

Keep digging to a minimum. While it is true that some loosening of the soil around a plant can be beneficial by opening up more volume for the roots to exploit, too much digging brings dormant weed seeds to the surface, upsets capillary connections, and, if done in two-wet soil, may even ruin the soil's natural structure. It is also very hard work.

Except for the actual planting of new trees and shrubs, you will be surprised at how little digging you can get away with. If you allow enough time, for the process cannot be hurried, you can even create a new garden bed without spading the soil. In the fall, mow the area closely and use an edging tool to delineate the shape. (Be sure to contour any curves to fit the radius of turn of your mower so that you can keep future trimming to a minimum.) Then lay a few layers of newspaper over the grass. Cover the newspaper with a few inches of leaves, straw, and/or manure and cover that with about two inches of compost. Let the whole thing rot over the winter, and it will be ready to plant in the spring.

Virtually all weeds are plants of bright sunlight, so always keep your soil well shaded. This can be done with a good cover of mulch, such as leaf litter, which will also conserve water and make the weeds that do appear far easier to remove. You can also position your plants so that, when mature, their leaves overlap slightly. Obviously this result can be far more quickly achieved with herbaceous plants than woody ones, so underplant young trees and shrubs with groundcover or use temporary plantings of meadow flowers and native grasses that can be removed or exchanged as the trees grow and the shade deepens. The edge of any garden bed is the hardest spot to maintain, since the weeds tend to move in from the lawn and vice versa. If you install a metal edging strip, use the heaviest and most permanent you can find. A mowing strip, that is, a six-inch-wide paving of bricks or stone set flush with the grass, is even better and will make the job of edging much easier.

Whenever possible, get the weeds out while they are small, and before they go to seed. One year's seeding can result in many years' weeding. And, since the very act of weeding also disturbs the soil, always be sure to replace the cover of mulch.

There are some weeds, however, that spread by underground stems, the smallest piece of which will result in a vigorous new plant. Attempting to pull these weeds out by hand is a waste of time and labor, but they can be eliminated with the judicious use of herbicide. The very thought of using a herbicide alarms many people, yet there are some, such as glyphosate (found in Roundup and Kleenup), that are quickly tied up in the soil or in the plant and break down into harmless compounds in a matter of weeks. Nevertheless, because glyphosate, like most herbicides, should NOT be used near water, NEVER dump leftover herbicide down the drain. Always follow the directions to the letter, and USE UP the quantity you have mixed.

Glyphosate is also nonselective, which means it will kill both dicots (broad-leaved plants) and monocots (grass and other strap-leaved plants), so it must be applied carefully and precisely to avoid damaging neighboring plants. There are various techniques to achieve this. In a garden bed or small meadow, you can pull a cotton glove on *over* a rubber one, dip your gloved fingers into the herbicide solution and wipe the mixture directly on the leaves. Or make a gel mixture (see recipe at the end of this chapter) and dribble it onto the plant from a plastic squeeze bottle. You can also use a roll-on applicator bottle such as many deodorants are sold in.

Forget about using any kind of synthetic insecticide or fungicide, however. They kill all kinds of beneficial creatures like bees, solitary wasps, butterfly larvae, and earthworms. Nor are they good for children or pets, not to mention songbirds, squirrels, rabbits, and a host of microscopic organisms. In subtle ways, these compounds alter the complex ecosystem of the garden, unbalancing the vital networks that normally hold insect pests and plant diseases in check. In the long run, using them will only make things worse. A recent study done by Cornell University reported that in spite of the fact that farmers now use thirty-three times the amount of pesticides that were used in the 1940s, and the chemicals used are ten times more potent, 37 percent of all cultivated crops is now lost to insects, versus 31 percent in the 1940s. Further, the report states that, in 1945, when corn was regularly rotated from field to field and no insecticides were used, only 3.5 percent was lost to insects. Now farmers who depend only on pesticides and do not rotate their corn lose 12 percent of their crop.

The fact is that most plants, unless they are under some kind of stress, can take care of themselves. If you maintain acceptable levels of organic matter in your soil, choose plants that are naturally suited to your site, mulch to reduce stress from drought and extreme temperatures, and prune to increase air circulation, you will find little need for any kind of pest control.

If a problem does appear, there are a few easy things you can do, such as: hosing down the infected plant (water sprays can often remove insects), cutting out a diseased limb, or even replacing the entire plant on the grounds it is not suited to your environment. Beyond that, Japanese beetles can be dealt with by spreading milky spore disease, which is now available in granular form, to kill the grubs in your lawn. (Unfortunately, to be most effective your neighbors must use it, too.) It takes two to three years to complete the job, but you will then be free of grubs for up to fifteen. Sparrows also eat Japanese beetles, and toads eat slugs. Predatory wasps, ladybugs, and green lacewings all

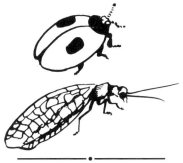

LADYBUG, FAMILY COCCINELLIDAE
GREEN LACEWING, *Crysopa*, spp.

feed on aphids, although it is futile to try and introduce these insects. If you have a juicy crop of aphids, and have used no poisons to deter these predators, they will arrive on their own.

Baking soda will control powdery mildew, blackspot, and other fungal problems, as well as all kinds of scale. Mix it at the rate of one tablespoon per gallon of water and add half a teaspoon of liquid detergent and a few drops of cooking oil to help the mixture adhere to the leaves. White flies can also be treated with a mix of equal parts of water, ammonia, and rubbing alcohol. To avoid spreading any kind of problem, always disinfect your hand pruners.

Take the trouble to learn as much as you can about the problem that is plaguing you. It is particularly important to know the life cycle of an insect pest so you can deal with it at the appropriate time. Keep an insect guide handy so you can tell the good ones from the bad ones. Temperature also affects insect development, so keeping a record of daily minimum and maximum temperatures can tell you when certain insects may begin to appear. You can also monitor a particular insect's presence with traps like sticky cards (yellow for white flies, blue for thrips) or pheromone traps that attract them with hormones. Remember, the idea is to control the problem, not necessarily eradicate it. Learn to live with a small amount of damage to your plants.

If you keep your soil well supplied with organic mulch, you will rarely need to water or fertilize your garden. In fact, using too much nitrogen fertilizer can actually make your plants more susceptible to disease and drought as well as shorten their life. If you must feed the lawn, use organic nutrients like Lawn Restore that nourish the grass slowly over a long period. Synthetic fertilizers tend to promote short-lived growth spurts and, if they should wash into ponds and streams, can cause explosive growths of algae that rob the water of oxygen, killing fish and other organisms. Synthetic lawn fertilizers also promote thatch, a barrier of dead grass that blocks penetration by water and nutrients and must be raked or otherwise removed. Organic compounds include microbes that break down dead grass, thus improving the soil and reducing your labor at the same time.

Never send leaves to the dump. Compost them instead, or use them as mulch. Whole leaves can be used as they fall around woodland trees and shrubs, but it is better to chop them by running a lawn mower repeatedly through the pile before putting them around smaller plants and groundcovers. There is also a large-wheeled vacuum on the market that both chops and bags leaves so they can be stored for later use. In either case, it is better to wait until the ground is frozen before returning the shredded leaf litter to garden beds to avoid frost heaves. By then the mice should have found other homes as well.

In a small city garden, cover your plants in September with netting (the green sort used to protect berry bushes from birds) and pick up your leaves all at once.

You will reduce your maintenance considerably if you plan your garden in a way that reduces or eliminates endlessly repeated tasks. We have already mentioned ways to reduce weeding. Mowing time, too, can be reduced if you replace part of the lawn with shrubs and groundcover. Establishing the new planting will certainly take more time in the beginning, but in the long run it will take far less and be far more attractive, artistically as well as ecologically.

Natural gardeners always treat their gardens with the respect they deserve as living links to the wider natural world. They let their home region and homesite dictate their choice of plants, allowing each one ample room to grow and develop. They shape and prune gently and naturally, enhancing each plant's inherent beauty and structure, as well as its role in the overall design. They rely on natural processes to nourish and replenish their soil. They make use of nature's own strategies to discourage invasive weeds, insect pests, and disease. Finally, they welcome life into their gardens, celebrating it in all its interesting forms and relationships.

> **ROUNDUP GEL RECIPE (COURTESY OF MT. CUBA CENTER) GREENVILLE, DE**
>
> 1 LITER WATER
> 40 MILLILITERS ROUNDUP
> 25–30 GRAMS GEL POWDER (MILLER LIQUA-GEL)
>
> 1. Place water and Roundup in blender.
> 2. Mix at lowest speed.
> 3. Slowly add gel powder while mixing on lowest speed.
> 4. Cover and increase speed to high until the consistency of mustard.

·T·W·E·L·V·E·

Releasing the Native Landscape

However sensitively and skillfully done, no private garden or cultivated landscape can ever approach the beauty and complexity of an unspoiled natural area. Such places truly hold the genetic keys to the preservation of the world. To counteract the relentless pressures of development, we need to protect such places whenever and wherever we can. For the best of them, the skills needed may be more political than botanical, for no pristine community can ever be considered safe unless it is legally protected. Yet, even when that is accomplished, these areas will still need watching to guard against the subtle but degrading effects of human use and nearby development.

The open space we are most concerned with here, however, are those woods and open fields that, to one degree or another, have already been invaded by alien weeds. Because both the idea of restoring such areas, as well as the best techniques to use, are new to so many of us, this chapter will include tips from a variety of sources in the hope that you will choose the method or combination of methods that works best for your particular site.

Whether you are attempting to reclaim part of your own property, or are responsible for the management of a protected open space, such as a park or community nature center, the problems will be the same. The scale, however, may be quite different. Obviously, managing a large open field of thirty acres or more will require equipment and

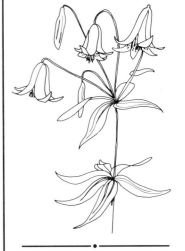

CANADA LILY
Lilium canadense

labor far beyond what is required for a small suburban meadow. When it comes to woods, however, the opposite is true. The dense shade of an unbroken swath of mature forest will discourage invasion by most exotic weeds, whereas the high proportion of sunlit forest edge makes any wooded fragment of less than five acres continually vulnerable.

Why must we meddle at all in an already natural landscape? Why not simply let nature take its course? Unfortunately, while we can all recognize a weed-choked garden, comparatively few of us seem to notice the destructive impact of invasive aliens on the natural landscape. Some even argue that these escaped imports enrich the countryside in the same way that our country is enriched by people from other cultures. Unfortunately, this is rarely the case. Few of these plants contribute to natural diversity. Instead, like a conquering army, they can degrade a plant community to such an extent that the number of species found there may quickly drop to a mere handful.

The weeds that plague our woods and fields are not, for the most part, the same ruderals that appear in the garden, but imported vines, shrubs, and trees that, assisted by birds and other wildlife, are able to spread unhindered by the viruses, pathogens, or predators of their native habitats. Unfortunately, the ability of these plants to capture all the available space is so powerful and persistent that many land managers resort to frequent and widespread mowing as their only recourse short of abject surrender.

Nevertheless, there are some who are learning to manage land in a more ecologically constructive way. By restoring the grandeur of our remaining woods, and maintaining open fields in native grasses and wildflowers, these people are also reviving some of the lost habitat that so much of our wildlife depends on. Their efforts are part of a worldwide movement that seeks not only to arrest further degradation of the natural environment but to restore it to at least a semblance of its former condition. To this end, they are replanting prairies and savannas, anchoring dunes, and reclaiming marshes and other wetlands all over the planet. In them lies the hope of the world.

While these restorationists employ some gardening techniques, they are not gardeners, but naturalists. The primary goal of their efforts is not to create natural beauty, but to preserve biodiversity. That is, to once again allow all the original plants, mammals, reptiles, birds, insects, spiders, fungi, earthworms, and hosts of unseen microorganisms that are found in any natural area to thrive and function normally. Rather than imposing order and beauty from without, according to their own tastes and desires, they permit it to appear from within, as a by-product of nature's own intrinsic dynamics.

Most restorationists are trained scientists who work on such large-scale and daunting projects as strip mines, landfills, and phragmites-infested marshes, whereas we are primarily amateurs concerned with releasing a degraded woods or field. Nevertheless, the complexity of the problems, as well as their solutions, are similar. In the end, the measure of our success, like theirs, will be determined by how well the restored community is able to continue functioning on its own. The good news is that, given the time and opportunity, many restored native communities will be able to do that with a minimum of ongoing maintenance. The bad news is that we will never again be completely free of the alien plants that threaten them.

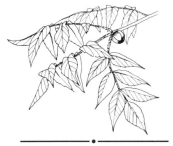

MOCKERNUT HICKORY
Carya tomentosa

WOODS

Forests, in all their variety, are the East's only stable upland plant community. With the exception of coastal dunes and marshes, barrens, and northern bogs, virtually all the land east of the Mississippi was originally covered with trees. Today, with a few rare exceptions, nothing remains of that original forest; virtually all the woods that now exist have either been logged repeatedly over the years or are growing on previously cleared lands. Few of them contain trees more than seventy-five years old.

Nevertheless, we do have some forests that are filled with majestic oaks and hickories, huge beech trees, red and sugar maples, mature white pines and hemlocks, and a host of other species, depending on the region of the country. A common characteristic of these older woods is an open and clear delineation of their various layers. That is to say, the difference between canopy trees, understory trees, shrubs, and groundcover of wildflowers and ferns is obvious even to the most inexperienced eye. There is also a feeling of openness; in spite of the thick shade and tall trees, one can walk through these woods with ease.

Unfortunately, this is seldom the case in any woodland less than forty years old. All too many of these younger woods are characterized by smothered and broken trees so covered with entangling vines that they form an impenetrable wall of vegetation. Even the largest trees in a degraded woods may be distorted by encircling bittersweet, and the shrub and sapling layer may be discernible only as mounds in a blanket of honeysuckle. There is simply no space for wildflowers or ferns.

To release such a woodland can be a slow and difficult task. It is better to resist the urge to clear it all out at once, but rather to begin in the center, where the trees are largest and the shade deepest, and work slowly outward toward the edge. The idea is to move from the best *toward* the worst, never the other way around, so as to allow existing rootstocks and seeds of native plants the best chance to regenerate. If you start, as many of us do, at the outer edge where the problem is at its most severe and more inward toward the healthier parts, you will not only let in more light, thus further stimulating the growth of the vines, but the majority of regenerating rootstocks and seeds will be those of the very aliens you are trying to defeat.

Whatever your method of clearing, try to avoid leaving bare and disturbed soil since it, too, will favor aliens over natives. Any healthy woodland floor is covered with a litter of leaves, twigs, branches, and fallen logs in all stages of decay. Not only does this natural mulch contain all sorts of natural organisms that are essential to the health of the forest, it also insulates the soil, holds moisture, and prevents erosion. Best of all, it tends to inhibit the seedlings of many alien weeds.

The older the woods, the easier time you will have to restore it. However, even an old woods, particularly in the Piedmont section of the country, may have invading Norway maples and garlic mustard, *Alliaria petiolata*, a herbaceous plant that is fast replacing native wildflowers along some woodland stream corridors. While garlic mustard is quick to spread across a woodland floor, its shallow roots make it easy to pull out. If done in the spring when there are no ripe seeds to worry about, the uprooted weeds can simply be added to the forest litter.

The maples are quite another problem. Both the Norway and the sycamore maple have very large, thick leaves which emerge early in the spring, creating a deep shade that severely inhibits native woodland wildflowers. Both trees are also strongly allelopathic: their roots release a toxic substance into the soil that suppresses the growth of competing species. Unless controlled, therefore, they will eventually dominate the woods.

Neither the Norway nor sycamore maple turns yellow until mid-November when most other woodland trees are bare. In October, their leaves show a conspicuous deep green among the flamboyant reds and yellows making this a good time to identify trees for removal. To be entirely sure, break off a leaf. If the petiole exudes a milky sap, the tree is definitely a Norway maple.

∴ FIELD AND FOREST ∴

Seedlings and small saplings should be pulled up wherever they are encountered. (A Weed Wrench is a good tool for pulling out saplings up to two inches in diameter.) Young trees can be cut down and the stumps painted with herbicide. Large mature trees can also be girdled, a job that is most easily done in the spring. However, girdling alone will do nothing to deter a Norway maple; unless you paint the wound with herbicide, it will be soon calloused over and the tree will go on as before.

Mixing the herbicide with oil (kerosene, diesel, or fuel oil) is far more effective than mixing it with water, which tends to run off the tree trunks like rain. Using oil as the base also makes it easier to tell which plants have already been treated.

You may also find plants of privet, bush honeysuckle, or multiflora rose, even in the middle of a relatively mature woods. These, too, can be pulled out with a Weed Wrench, or cut and the stumps painted with herbicide. There will surely be Oriental bittersweet vines; here with thick stems that twine up the forest trees like huge tropical snakes. If you simply cut them at the base and paint the stump to prevent resprouting, the top will eventually rot and fall out of the tree. Mature honeysuckle vines can be treated in the same manner, although honeysuckle on the forest floor that has already been somewhat weakened by the shade can usually be pulled up fairly easily.

When treating a stump, cut it close to the ground and be sure to cover the outside bark and cambium layer as well as the root collar at the soil line. Just painting the cut surface will have little or no effect. Many trees and shrubs can be more easily treated in winter, when the foliage is off and the insects are gone. However, vines, such as honeysuckle, porcelain berry, and Oriental bittersweet, that quickly resprout from cut stems or roots, are best treated in the late summer and early fall, when the sap is moving down into the plant's root system.

RELEASING THE NATIVE LANDSCAPE

Again, begin where the cover is sparse, and work into the more severe infestations. Where the growth is especially dense, you can apply herbicide with a sponge, if you are careful to first remove the vines from desirable trees and shrubs. Wear rubber gloves to protect your hands. You can also use a newspaper or a cardboard shield to protect the bark of trees. Be patient: herbicides work so slowly that you may not notice any effect for weeks, perhaps not until the next growing season.

Except for alien maple seedlings and garlic mustard, the deep shade in the center of a large woods usually discourages the reappearance of most alien shrubs and vines. They will, however, continue to invade forest clearings and any place where the canopy is thin or broken. As you move toward the edge of the woods, however, the infestation will become more and more severe, until there may be little else but a dense, smothering tangle of alien shrubs and vines at the forest edge. Any woodland edge, in fact, is a battleground between the forces of sunlight and shade. In those place where the strangling vines have succeeded in breaking the trees and thinning the cover, all kinds of invasive shrubs will thrive; only under the largest trees will the shade be able to weaken their hold.

Typically, the largest trees at the edge of the woods are left over from an earlier stage of succession. Many are large black cherries, *Prunus serotina*, that stand strangled and broken by bittersweet, honeysuckle, or porcelain berry, amid a dense growth of multiflora rose, privet, or bush honeysuckle. This pattern is often found along old fencerows as well.

Nevertheless, a careful examination of any woodland edge or fencerow will also reveal native trees from the next successionary wave; depending on the area, these might be tulip poplar, dogwood, redbud, red or sugar maple, white ash, sassafras, or birch. There may also be pines, red cedars, and shrubs such as shadbush, certain species of sumac

TRUMPET VINE
Campsis radicans

and viburnum, or blueberry and huckleberry. There will almost certainly be brambles such as greenbrier, *Smilax* spp., blackberry or black raspberries, as well as vines like trumpet vine, *Campsis radicans*, or poison ivy.

One way to attack the edge problem is to keep certain selected trees clean of vines, and simply allow them to grow and flourish, gradually increasing the shade around them. Using such a tree as the reference point, you can also clear a pie-shaped area behind it. A row of several such triangles can then be gradually widened until they eventually meet behind a narrow rim of alien shrubs and vines at the very edge of open land. The idea here is to avoid a sudden increase in the light penetration by allowing a healthy growth of released natives to become established before removing the last of the weeds. Trees grow at their own rate, however, so the effect of such a method may not be obvious for several years. The final result, however, should be a more stable edge.

Nevertheless, birds will continue to drop seeds of these plants, and the old roots and stems, particularly of bittersweet and porcelain berry vine, will regenerate unless they are killed with herbicide. To help keep any undesirable seedlings from gaining a foothold, it helps to limb up, or cut off the lower limbs of, the outermost trees and shrubs so a mower can get close to the trunks.

In the case of a hedgerow that has been taken over completely by alien vines and multiflora rose, the only recourse may be to cut down everything but the largest and best trees and monitor what comes back, carefully spot-treating the aliens with herbicide while allowing the natives to grow up. The light penetration in any hedgerow makes them especially difficult to manage. Even if you eventually succeed in killing all the old roots (not an easy job), it will be virtually impossible to keep new bird-planted seeds from sprouting.

If your patch of woods is less than five acres, consider

enlarging it by letting new trees develop around the edge. Or combine two or more fragments into a larger whole by allowing young trees to grow up between them. Restoring and connecting our existing woodlands so that they form corridors of natural habitat is one of the most valuable things we can do to halt the alarming decline of migratory songbirds. Warblers and other woodland birds, such as thrushes and tanagers, need the protection of the deep woods to keep their eggs and young safe from marauding cowbirds and blue jays. Studies have also shown that, once they have reached a nesting territory, many woodland birds will not cross open land. Therefore, small isolated patches of woods make it almost impossible for them to find mates and raise their young successfully.

BLACKBERRY
Rubus, spp.

FIELDS

In the East, every open field represents a particular stage of plant succession. Each, by its very nature, is a temporary community that must be managed in order to survive. Nor are there many eastern fields without at least some alien forage grasses and imported field flowers. Many of these not only are attractive in their own right but also provide valuable food for wildlife, especially butterflies and other insects. One exception is the Canada thistle, a plant that spreads widely from underground roots and, if not controlled, will form a prickly blanket covering many square yards. Other kinds of thistles are biennials and cause little trouble; they are also a great attraction for goldfinches and butterflies.

There is no doubt that fields can be among the most interesting and rewarding places to work with the natural landscape. A grassy field interrupted by swaths of wildflowers and dotted with islands of shrubs and trees not only

∴ FIELD AND FOREST ∴

is beautiful but is also a prime wildlife habitat. Rabbits as well as birds nest in fields, butterflies flit among the flowers, and hawks and foxes hunt grasshoppers and voles in the grass. At the same time, the combination of disturbed ground and bright sunlight makes some fields so vulnerable to invasion by alien shrubs and vines that a mere two or three years of neglect can turn them into an impenetrable tangle of brush.

It is evident to anyone familiar with the rural landscape that some fields are far more vulnerable to this sort of invasion than others. The reasons for this are complex and sometimes conflicting. They include the underlying geology (fields on metamorphic rock seem to be more vulnerable than those on sedimentary rock) as well as the way in which the field has been used in the past.

Most of the plants, both native and alien, that appear first in an abandoned field come from roots and seeds already in the ground. Therefore, a field that has been under active cultivation for many years will tend to have fewer viable rootstocks than one that has been previously used for hay or pasture. At the same time, a rich fertile farm soil that has been regularly limed will significantly assist the invasion, whereas a soil that is both thin and exhausted may only be able to support swaths of broomsedge and red cedar.

Therefore, it is important to analyze both the particular characteristics and the present condition of your site before deciding on a management strategy. Basically, the choices are: one, to keep all of it in grass and herbaceous plants with the exception of a few isolated trees; two, to keep most of it open but allow one or two islands of shrubs and early successionary trees to develop (to maintain a certain scale and density, you will also have to remove trees of the next stage of succession); and three, to permit, or even assist, the natural return of woods.

As with any kind of natural landscape, you must begin

by taking an inventory of the plants already growing in your field. If you are lucky enough to have a field full of native grasses like broomsedge, little bluestem, and Indian grass, *Sorghastrum nutans*, interspersed with various kinds of goldenrod and asters and dotted perhaps with isolated groups of sumac, blueberry, or blackberry, as well as early successional trees like red cedar, sassafras, or hawthorne, your task will be relatively simple. A high-quality old field such as this needs only to be mowed annually.

To protect nesting birds, this should be done in November or during the first two weeks of March. The latter is preferable because it provides ample winter cover for wildlife and maintains a high insect population for fall migrating birds, particularly kestrels and other hawks. A November mowing, on the other hand, will distribute the seeds of wildflowers more effectively. If you have little or no problem with exotic weeds, you can try mowing every other year to increase the incidence of biennial and perennial flowers. Be sure to mow around any trees and shrubs you want to keep. Isolated groups of woody plants, such as dogwood, sumac, blueberry, or young sassafras will increase not only the visual interest of your field but its value to wildlife as well.

MULTIFLORA ROSE
Rosa multiflora

Fields of Andropogon and goldenrod apparently have such a strong allelopathic effect on other plants that they can maintain an old field's character for as long as thirty years. Old pastures and hayfields that were originally planted in European forage grasses, however, lack this attribute and may have scores of small plants of multiflora rose, honeysuckle, Oriental bittersweet, porcelain berry vine, or other exotics hiding in the grass. Mowing alone does little to discourage these plants, particularly porcelain berry vine and multiflora rose, both of which will lie in wait for years for a chance to grow up. You must kill their roots in order to eliminate them.

Mowing ten times a summer for five years is one way

to accomplish this, but then, of course, you will have a large sterile lawn with little visual interest, and no birds or insects. A better way might be to mow once or twice a year to keep the exotics under control, while you gradually spot-treat the invading plants with herbicide.

Use a brush hog mower or sickle bar mower set at four to six inches. The first mowing should be in March, as above, and the second during the first two weeks in August, after most birds have finished nesting and before any exotic plants have gone to seed. A midsummer mowing can attractively reduce the height of goldenrod while allowing enough time for some wildlife cover to return before winter. There will, however, be some adverse effect on butterflies, which depend on August flowers for nectar, and to the late-nesting American goldfinch, which depends on the seed of thistles and other plants to feed its young.

Allow the field to develop slowly, keeping it well under control while you slowly eliminate the invasive aliens. As time goes on, you may want to introduce seedlings of native grasses and forbs as well as create islands of native shrubs and early successional trees, either by allowing already present rootstocks to develop or by planting out young seedlings. Treat these thickets like a forest edge; that is, limb up the trees and shrubs, so you can mow underneath the outer branches to keep them free of vines.

If your field already has a good growth of trees and shrubs throughout, both native and exotic, your best bet may be to accelerate the natural thrust of succession. Speeding the return of woods takes less time and labor in the long run; moreover, if it also enlarges an already existing woodland or restores a degraded stream valley, it may be the most important thing you can do to help our declining populations of woodland birds.

The first step is to take a careful inventory of the species already growing in the field. Unfortunately, a badly degraded field will have far more exotics than natives. If yours is a mass of multiflora rose, Japanese honeysuckle, various species of bush honeysuckle, porcelain berry vine, and perhaps even swaths of Canada thistle, it will rarely succeed to anything. In that case, the best approach may be to cut the whole thing down with a heavy-duty brush hog mower, wait a few weeks, and treat any resprouting exotic weeds carefully with a selective herbicide. All those that contain 2,4D will kill broadleaf plants but not grass. Again, the best time to use them is in late summer and early fall, when the carbohydrates are moving into the plant's roots for winter. You may have to repeat the process again the following year. At first, it will also take constant care and vigilance to keep reinvading vines out of a developing grove of young trees.

If on the other hand, natives already outnumber the exotics in your field, work your way gradually across it, marking the plants you want to save and eliminating the rest. You can do this in various ways. One way to get rid of multiflora rose, as well as other twiggy shrubs, is to back a brush hog mower into the center and simply grind it up. The same effect can be accomplished on a smaller scale with a gasoline-powered hedge clipper. The shredded twigs can be left on the ground as mulch.

In a few weeks, go back and treat the resprouting stump with herbicide. Choose the least persistent one that will do the job, mix it precisely according to directions, and saturate the bark at the base of a shrub to about eighteen inches from the ground. It is never a good idea to spray a herbicide, since the smallest movement of air may cause injury to the very plants you are trying to encourage. Instead, open the nozzle so that the chemical is dribbled rather than sprayed on the target. Remember, *never* use a herbicide close to a stream or spring.

To speed the growth of trees and discourage reinvasion, you can also shade the plots with shadecloth and treat the soil with sulphur to increase the acidity. Sulphur also removes excess phosphorus and releases aluminum and toxic metals.

Fire as a management tool is an alarming idea to many people, but the controlled burning of meadows, shrub thickets, and even woods is a technique that is being used more and more by professionals. Although the prairie ecosystem is dependent on fire for its very existence, this is not generally the case in the more humid East. Nevertheless, we can use fire to manage open fields. There is evidence that it favors plants such as the wild indigo, *Baptisia tinctoria*, huckleberry, and sweet fern, *Comptonia peregrina*. A program of annual or biennial burning at the Connecticut Arboretum has also increased the height and vigor of many grasses, sedges, and goldenrods.

Fire has also been used to manage some woods. Again, controlled burning at the Connecticut Arboretum over the last twenty years shows that when accumulated litter on the forest floor is burned, it causes a dramatic increase in the number of oak seedlings as well as the incidence of some herbaceous plants, notably spotted wintergreen, *Chimaphila maculata*. Trees such as dogwood, black cherries, and hickories readily resprout after a fire, although river birches, *Betula nigra*, do not. Both the native and exotic species of maples tend to be killed as well.

Most communities, of course, prohibit the setting of open fires without a permit. Such permits usually require that the local fire department is present and in charge, and some departments use such controlled burns as training sessions for their members. Burning is usually done in early spring, although some studies are now being done on the effects of summer and winter fires as well. It is surprising how quickly the blackened ground springs into active growth.

Anyone who works to release and restore any kind of natural landscape finds it a humbling as well as rewarding experience. The strength and persistence of the exotic invaders that we have so carelessly let loose into the environment provide a powerful lesson in the consequences of human ignorance and heedlessness. At the same time, the strength and resilience of a released native community can furnish an equally powerful lesson; this time in the marvelous fecundity and creative power of nature.

What greater satisfaction can there be than to look out on a field of native grasses and flowers that is alive with butterflies and birds, or walk through the solemn silence of a majestic woodland, and know that we have actually participated in its renewal. The best lesson of all may be the feeling that our human presence in nature does not have to be destructive. We can heal as effectively as we can harm. We can, in fact, become the agents of hope.

NEW ENGLAND ASTER
Aster novae-angliae

·T·H·I·R·T·E·E·N·
Acquiring Native Plants

CAROLINA SILVERBELL
Halesia carolina

It wasn't so long ago that native plant gardeners, as well as native plant nurseries, dug from the wild as a matter of course. Old books on woodland gardening invariably included chapters on when and how to "collect" plants as well as how to care for them in the garden. Nowadays, such behavior is considered no better than stealing. The enlightened gardener either propagates his or her own, surely one of the most enjoyable aspects of gardening, or is careful to buy only from nurseries that *advertise* that they propagate the plants they sell.

Fortunately, the number of ecologically responsible nurseries has grown considerably in recent years. These growers are careful to avoid any depletion of natural populations by removing only seeds and cuttings from the wild. However, because many older, more traditional nurseries still follow the old ways, no gardener can ever assume that a plant has been nursery propagated unless that fact is clearly stated on the label or in the catalog. Be careful, too, of the terms used; the phrase "nursery grown" may mean only that the plant has been on the premises for more than one season.

Native plant nurseries acquire plants in three ways: they propagate them from seeds and stock plants obtained from a variety of sources; they dig plants from the wild and sell them "as is"; or they buy plants from wholesalers, who may or may not have collected them from the wild. Thus

∴ ACQUIRING NATIVE PLANTS ∴

even a nurseryman who does not "wild collect" may buy from suppliers who do. Many of these so-called suppliers turn out to be ordinary residents in choice locations who earn money by digging wild plants. It is hardly surprising that some of the best sites are soon depleted, and the collectors forced to move farther and farther afield.

Some nurserymen insist that removing a few plants here and there of an abundant species does no permanent harm, and, if the collector has enough control over the site to regulate the numbers that are taken, they may be right. This, however, is rarely the case, and even "a few plants here and there" can soon reach alarming proportions. It has been estimated that well over one hundred thousand ornamental plants are removed from the mountains of North Carolina and Tennessee every year, a rate that has caused even the most abundant species to become scarce in some areas. It is now illegal in North Carolina to trade in certain species unless the plants are certified to be horticulturally propagated. Eventually, it is hoped that every nursery dealing in native plants will come to abide by similar propagation requirements.

Until that time, however, gardeners should be mindful that the recent surge of interest in native plants could exacerbate the problem, since the number of species plants collected from the wild rises, and falls, according to the demand gardeners place on the nurseries. The pressure to fill more and more orders could force a grower to collect and, inevitably, have a negative impact on some wild populations.

Every wild population is important, even of a relatively common species, because each contains a diversified gene pool: a sort of genetic bank account of adaptation potential. It is this hidden reservoir of capacity for change that allows a species to survive unexpected alterations to its environment. However, if the number of individuals in any one population falls below a critical number (known as the genetic minimum), further decline becomes inevitable. Clearly, every native plant gardener must learn to make intelligent choices, based on a thorough knowledge of the plants themselves, in order to avoid buying any that may have been dug from the wild.

Of all our native flora, woodland wildflowers are at the most serious risk; particularly the monocots (flowering plants that generally have flower parts in multiples of three, straplike leaves, and parallel veins) because they are more difficult for nurseries to propagate than dicots. In fact, the woodland species deemed most vulnerable to digging for commercial sale are members of two monocot families: the orchids, Orchidaceae, and the lilies, Liliaceae. Members of the iris family, or Iridaceae, are also extremely vulnerable.

Native terrestrial orchids are especially threatened. For this reason, botanists agree that inexperienced gardeners should not attempt to grow these plants. Terrestrial orchids have never been propagated successfully in the quantities necessary for commercial trade; therefore, virtually every plant offered for sale has been dug from the wild. No responsible gardener should ever buy one of these plants. Not only will such a purchase contribute to the unnecessary destruction of the species in the wild; the collected plant will rarely survive for more than one season. Terrestrial orchids depend on the presence of a group of fungi called *mycorrhizae*, which live symbiotically on the plants' roots. For any orchid to prosper, both its needs and those of the fungi must be met—an extremely difficult assignment for any gardener. While a considerable amount of research is now being done on propagating terrestrial orchids, supplies of propagated plants are, as yet, far too small for commercial sale.

Many states now have laws forbidding the poaching of wild plants from public or private land, or the transporting of endangered plants across state lines. In spite of this fact, orchids, particularly lady slippers, continue to be advertised widely. A grower in Michigan was recently fined after it was discovered that he was paying collectors seven to ten cents a plant for lady slippers, which he then sold to nurseries all over the East at twenty-five to thirty-five cents each.

Lilies are also threatened by collectors. The lily family includes such popular woodland plants as *Trillium*, trout lilies, *Erythronium americanum*, Solomon's seal, *Polygonatum biflorum*, bellworts, *Uvularia* spp., and wild hyacinth, *Camassia scilloides*. Many of these plants are dug in huge quantities because they are extremely slow to propagate from seed. The seed of white trillium, *T. grandiflorum*, for instance, takes at least two years to germinate and up to ten years to bloom. Obviously, it makes little economic sense for a nursery to grow it from seed. Mature plants can be divided, of course, but division is also far too slow to supply the quantities demanded by the commercial market.

Unfortunately, many nurseries dig even those woodland species that are easy to propagate, because many of these plants produce relatively few seeds and also require complex treatment to germinate. All too often, it is simply easier and quicker to take from the wild.

Certain native ferns are also vulnerable because they, too, are difficult and slow to propagate. Others are tougher, more widespread, and can easily be divided. Fern fanciers should know the difference before buying plants.

One way to avoid wild-collected plants is to buy only *named cultivars* that have been specifically developed for garden use. A named cultivar is a plant that has been individually selected, usually from a group of seedlings, for a particular characteristic. More and more of these cultivars are appearing in the trade, and many of them make wonderful garden plants. Because they occur rarely in the wild and do not come true from seed, they must be reproduced by division, cuttings, or the newer technique of tissue culture.

There is no doubt that tissue culture can increase the

TRILLIUM
Trillium grandiflorum

supply of a single cultivar dramatically. Selected cultivars of the mountain laurel, for example, are now common in the trade, whereas a few years ago laurel was being dug out of the Appalachian mountains by the truckload because its cuttings were so difficult to root.

Nevertheless, tissue culture also has its drawbacks. Today's market forces are such that when a successful cultivar becomes available in the vast quantity made possible by tissue culture, it may come to dominate the trade to the point that it is planted almost exclusively. Because, like any vegetative propagation, the technique reproduces each individual plant's exact genetic makeup, it can result in a cultivated population of genetically identical individuals, each with precisely the same resistance or susceptibility to insect pests or disease. For example, in the 1970s, a disease called the southern corn blight was able to devastate the nation's corn crop, precisely because so many of the hybrids planted during those years shared the same genetic heritage.

Something akin to this situation may have contributed to the effect that Dutch elm disease had on the magnificent American elms that once graced streets and towns all over the Northeast. Equally important, of course, was the conspicuous lack of *species* diversity; since the elm was the only kind of tree planted, not only was the disease easily carried from one to the next by the elm bark beetle, it was also able to travel underground through intertwining root systems.

When the range of a native species in the wild is spread over many latitudes, it can also be important to know the *provenance*, or the regional source of the particular plant you are buying. We tend to forget that in nature, where seed is the primary method of reproduction, not only is each individual different from any other, but each population is different as well. For example, in a species like the red maple, which grows in a wide variety of habitats all the way from Canada to Florida, a tree grown from seeds or cuttings

∴ ACQUIRING NATIVE PLANTS ∴

originally collected in Ontario will have quite a different capacity for climatic survival than one that originated in the Everglades.

The fact that woody plants are usually propagated vegetatively, however, does mean that they are far less likely to be dug from the wild than herbaceous ones. Those in the heath family, Ericaceae, are the exception. We have already mentioned the mountain laurel. Another ericaceous plant, the trailing arbutus, *Epigaea repens*, has been virtually eradicated from many of its former sites by collecting. Many native azaleas are also at risk even though they can be grown quite easily from seed, and wild dug plants rarely do well in the garden.

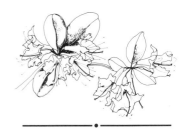

Rhododendron chapmanii

The point is, native plant gardeners should be careful to avoid any species that is known to be difficult to propagate. Nor should they ever buy a rare plant unless they are certain it has been propagated without the use of a wild population. Don't hesitate to ask where a particular plant came from and learn to recognize one that has been obviously collected. Any plant is suspect that has been recently potted, or that has leaves askew rather than symmetrically oriented as they would be in a nursery-propagated plant.

It is safer to get out of the woods and into the meadow. Most native grasses and meadow plants are easily propagated. Many nurseries now carry good supplies of seed as well as selected cultivars of meadow species like goldenrod, *Solidago*, Joe Pye weed, *Eupatorium*, black-eyed Susan, *Rudbeckia* and members of the aster family, Asteraceae. The pea family, Leguminosae, the snapdragon family, Scrophulariaceae, and the evening primrose family, Onagraceae, also include many species appropriate for natural landscapes.

Inevitably, the question of "plant rescue" or "salvage" comes up whenever gardeners talk about growing native plants. As the name implies, these are last-ditch efforts to save a population about to be destroyed by development.

FIELD AND FOREST

While permissible for the more common species (although even they might be better used for breeding stock or incorporated into nature trails), private gardeners should leave the rescue of rare plants to those experts who are trying to help the species to survive in the wild. No private garden in a country with as mobile a population as ours can be considered a safe haven for them, and if they are taken in the wrong way or at the wrong time of year the chances of success can be significantly reduced.

In the end, the only way we will truly protect America's rich botanical heritage will be to preserve natural habitats in sufficient quantity and diversity so that our native plant communities can continue to reproduce without disturbance. In the meantime, we can work to preserve a high level of biodiversity in our cultivated landscapes as well, by growing a large variety of both seedling plants and horticultural clones. Finally, the very best way to combat the dangers of indiscriminate collecting is to learn to propagate and increase these plants in our own gardens. The fun of gardening, after all, is not in the having, but in the doing and in the sharing: so be sure to exchange seeds and cuttings of your favorite plants with keen gardening friends.

IV

The Plants

·F·O·U·R·T·E·E·N·

What's in a Name?

The common names of plants are filled with a droll sense of imagery. Unsung country poets of an earlier day named them after their animals, like goosefoot or pussytoes, after their clothes, like monkshood or moccasin flower, or simply for their dreams, like forget-me-not or live-forever. Names such as poor man's weather glass or beggar's ticks show a charming self-deprecating humor. Hercules'-club and Venus' looking glass show a feel for classical myth, while St. Andrew's cross or Solomon's seal indicate at least a nodding acquaintance with religious history.

Country people listened to plants, they smelled them, they tasted and touched them; and they named them rattle-box, bittercress, skunk cabbage, or mouse-ear. They knew plants intimately, at all times of the day and at all seasons. For instance, St. Andrew's cross comes from the shape of the four-petaled flower, but it is the flat, polished seeds that form the "looking glass" of Venus, and the "seal" of Solomon only appears when the dead stem separates from the root.

These names are fun to know and use, but the truth is only a tiny fraction of the world's plants have common names at all, and most of those are only dry translations from the Latin. Greenish flowered pyrola, or panicled aster, are the kinds of names bestowed by botanists. They simply can't compare with the imagery of prince's feather or lizard's tail, yet prince's feather or lizard's tail will not do for accurate research. Such names are usually restricted to a particular region, or they may refer to quite different plants in different places. For instance, bush honeysuckle in the South means *Rhododendron*; in the North it is *Lonicera*. Fetterbush may mean either *Leucothoe* or *Pieris*, while bluebells can be *Mertensia*, *Polemonium*, or *Campanula*. Some common names can be frankly misleading: cypress spurge is no relation to cypress, and red cedar is not a cedar but a juniper.

Latin is the language of scholarship, and plants have been named and described in Latin for centuries, often in beautifully illustrated herbals. Because plants were medicine, proper identification was vital; each had to be described accurately and in minute detail. A widely quoted example from one of these early herbals is a description of clove pink, called "gillyflower" in early English gardens. It reads: "Dianthus floribus solitaris, squamis calycinis subovatis brevissimus, corollis crenatis," and translates as "Dianthus with solitary flowers, with very short inverted egg-shaped scaled calyces, and crenelated corollas." Detailed and accurate as this description is, it is undeniably awkward. Nor does it connect this dianthus to any other species or plant.

BOTANICAL NOMENCLATURE

Carl Linnaeus (1707–1778) not only loved plants, but was fascinated by their natural history and distribution. He wanted to give them names that would accurately describe a particular individual and at the same time connect it with other species in an orderly and expandable system. This in fact was his astonishing achievement. By the end of his life he had organized not only all the known plants, but the animals and minerals, too, in a way that made them accessible to scientists for the first time. Hundreds of thousands of new species have been discovered and named since his time, yet his system of nomenclature continues to work well.

Linnaeus was not the first to try and simplify unwieldy descriptions. The French botanist Charles L'Ecluse (Latinized to Carolus Clusius) was giving plants two names as early as 1576, but there was no consistent system to his effort. Linnaeus was the first to see clearly that there were groups of plants that had certain characteristics in common. He called these groups *genera*. Each group contained many different kinds of plants that while resembling each other in certain basic ways, differed in other ways that were consistently reproduced. He called these individuals *species*. In his book *Species Plantarum* he assigned each plant a number and described it first in the traditional way in Latin. He then took the genus name and combined it with a short name that he called the *nomen triviale*, or trivial name. Technically it is now called the specific epithet, although it is usually referred to as simply the Latin name. Ever since, these generic and specific names have been used together and comprise what botanists call the binomial system of nomenclature.

The First International Botanical Congress met in Paris in 1867 in an effort to formalize Linnaeus's system so that it could be used by botanists worldwide. This meeting began an often stormy process that eventually led to the adoption of The International Code of Botanical Nomenclature, a code that has now standardized the naming of plants all over the world.

The most important parts of this code, from the point of view of the amateur, are as follows:

1. Valid publication of all botanical names began in May 1753 with the publication of *Species Plantarum*. This

ruling formally acknowledged Linnaeus as the originator of the binomial system and validated all his original names.

2. Botanical names were decreed to be independent of zoological names. This rule gave scientists the freedom to use a particular generic name for animals as well as plants.

3. Each scientific name is to be followed by the name of the author. When correctly written, the name *Liriodendron tulipifera* should appear *Liriodendron tulipifera*, L., indicating that this plant was named by Linnaeus. If, as often happens, a later botanist should reassign the plant to a different genus, the name of the original author is retained in parenthesis, as in *Ipomoea purpurea* (L.) Roth. Including the name of the author is a sort of botanical bookkeeping. It increases accuracy, but because it is clumsy for everyday use, it is usually omitted from less technical books.

4. A herbarium specimen of the original plant found by the author is always the standard against which all subsequent plants of the same species are compared. It is the nomenclature type for that species.

5. Publishing a new name requires certain formalities, but once published a name can only be replaced if a previously published name is discovered for that species. For instance, the name for pinxterbloom was recently changed from *Rhododendron nudiflorum* to *R. periclymenoides* when the latter was found to date from an earlier time.

6. Family names are formed by adding -aceae to the stem of a prominent genus within the family. There are eight exceptions to this: family names that were in general use before the adoption of the rules. However, each exception was given an alternative name that complies with the code. The eight, with their alternate names, are as follows:

∴ WHAT'S IN A NAME? ∴

Compositae — Asteraceae	Composite or aster family
Cruciferae — Brassicaceae	Mustard family
Gramineae — Poaceae	Grass family
Guttiferae — Clusiaceae	St. Johnswort family
Labiatae — Lamiaceae	Mint family
Leguminosae — Fabaceae	Bean family
Palmae — Arecaceae	Palm family
Umbelliferae — Apiaceae	Parsley or carrot family

Older, more traditional botanists tend to stay with the old names, while younger botanists have often been taught the new ones exclusively.

Latin names, like common ones, tell their own stories, for they are full of fascinating allusions for those equipped to read them. The early naturalists, including Linnaeus, were fond of naming genera after fellow botanists or other people they admired. For instance, *Kalmia* was named for Peter Kalm and *Franklinia* for Benjamin Franklin.

Often generic names were taken from ancient Greek names, such as *Agrostis* for grass and *Fagus* for beech. Others are simply descriptive: *Sanguinaria*, the generic name for bloodroot, comes from the latin *sanguis*, meaning blood; *Daucus* is the Latin word for carrot; and *Chrysanthemum* comes from two Greek words: *chrysos* meaning gold and *anthos* meaning flower. some generic names allude to the medicinal use of the plant. Examples would be *Schrophularia* which was thought to cure scrofula, or *Salvia* which means simply "to heal."

Specific names are more apt to be genuine Latin as well as more descriptive of the particular plant. Many refer to a plant's distribution: the first North American plants to be named were usually called *americana*. In an effort to be more specific the early botanists then tended to call everything north of New York *canadensis*, everything from the Middle Atlantic states *pensylvanica* or *virginica*, and plants from the South *caroliniana*. These names are repeated over and over. As time went on, they became more specific, naming *Rhododendron catawbiense* for the Catawba River in South Carolina and *Franklinia alatamaha* for the Altamaha River in Georgia.

Specific names also provide clues to the plant's appearance. For instance, *latifolia* means wide leaf and *angustifolia* means narrow leaf. *Tomentosus* means hairy, *giganteus* large, and *cernuus* nodding. Names that refer to a plant's habit of growth include *scandens*, which means climbing, or *repens*, which means creeping. Still other names refer to a plant's habitat: *palustris* means marsh-loving, *pratense* means of the meadows, and *arenarius* means growing in sandy places.

The rules of Latin grammar are loosely followed; that is, generic names always have a masculine (-us), feminine (-a), or neuter (-um) ending, and the ending of the modifying specific name follows suit.

People are often uncertain as to how to pronounce Latin names. It is simpler than it seems, for most American botanists simply pronounce the words as if they were English. All syllables are pronounced and the accent is always on the next to last syllable or the one directly before it, that is, the penult or the antepenult. If the word has only two syllables, stress the first. The last syllable is never accented.

CLASSIFICATION

The first man to try and arrange plants into an orderly system was not Linnaeus, but Theophrastus (370–285 B.C.), a student of Aristotle. He grouped them by form, separating trees, shrubs, and herbs, and differentiating between annuals and perennials. He even described the various structures of flowers, noticing whether their petals were fused or separate and if their ovaries were inferior (below the point where the petals are attached) or superior (above the point of attachment).

After Theophrastus, no further effort was made at classification until a French botanist, Joseph Pitton Tournefort (1656–1708), published a book called *Rei Herbariae* in 1700. Tournefort was the first to describe plants from an objective interest as well as for their medicinal value. He described over ten thousand plants, many beautifully illustrated. Like Theophrastus, he grouped them by form, but he also recognized incipient genera such as oaks, peas, and willows, and he separated flowers that had petals from those that did not. Tournefort's system, although still rudimentary by our standards, was widely used in Europe for nearly a century.

An English botanist called John Ray (1628–1705) was a contemporary of Tournefort. He devised the first systems to be based on the structure of plants, that is, the characteristics of their fruits, flowers, and leaves, and he was also the first to recognize the difference between monocots and dicots. Although botanically superior, his system was overshadowed by Tournefort's and went largely unnoticed until long after his death.

Linnaeus had a passion for order and arrangement. His botanical system was numerical, based on the sexual organs of the flower. That is, he grouped plants primarily by the number of stamens, or male organs, dividing them further according to the stamens' length and structure. He also put plants with imperfect flowers together (those with only male or only female parts) and those with infertile flowers, or no sexual organs.

The value of Linnaeus's system was in its simplicity. Anyone who could recognize and count the parts of a flower could immediately identify and classify an unknown plant. However, while it showed the natural relationships of many plants, it tended to separate some that obviously belonged together, so it is not surprising that it was soon challenged.

Two early challengers were Bernard de Jussieu (1699–1756) and his nephew Antoine Laurent de Jussieu (1748–1836), both members of a prominent French botanical family. Their system, also based on the botanical structure of plants, was distinguished by the fact that it was the first to recognize thirteen classes of angiosperms, or plants with seeds contained in an ovary, categories that are still in use today.

All these early systems were based on the belief that a particular species never changed. At that time, all living things on earth were presumed to have been created in their present form in the space of a single moment far back in history. In the nineteenth century, Charles Darwin's theory of evolution as presented in *The Origin of Species* shattered this idea and in the process revolutionized all theories of classifying plants.

The basic tenet of Darwin's theory is that all species of plants or animals present on earth at this moment are descended from primitive ancestors. They attained their present form slowly, over thousands of generations, gradually becoming more and more adapted to a particular environment.

In plants, the tools of change and adaptation were hybridization and mutation. That is, today's plants are the result of hybridization between other, possibly quite different, species of long ago. Mutation—the term means a spontaneous change in the molecular structure of a gene that is inherited by the next generation—was probably responsible for abrupt changes. Of course, plants are still hybridizing and mutating today, but if the offspring so produced are not superior to their parents, they disappear, a fact which has undoubtedly been true from the beginning.

The theory of evolution spawned a whole new perspective on the classification of plants. So many new systems were developed incorporating the new theory that it is not surprising that no single one is used universally today. The differences between them usually center on what might have happened in the distant past that would indicate how large groups of plants are related. As a rule, botanists tend to be either "lumpers" or "splitters," disagreeing about what constitutes a separate species and what does not. This may be disappointing news to the layperson looking for a definitive and universal authority, but it does tend to humanize the scientist who, however intelligent and highly trained, is still just an observer in a complicated world.

Lest you assume that there are no points of agreement, however, we should add that certain broad classifications are universally accepted, and that botanists only dispute about how certain plants should be allocated among them. The categories of classification proceed from the general to the specific, dividing like the forking branch of a tree. They may be briefly summarized as follows:

Kingdom. Is the subject a plant or an animal? The answer may seem self-evident, yet it is not so obvious on the microscopic level, where some minute swimming creatures contain chlorophyll.

Division. Is it a vascular or nonvascular plant? Vascular plants have fluid-conducting cells, in stems, roots, and leaves. They include most of the conspicuous plants of the earth. Algae, fungi, mosses, and liverworts are nonvascular plants.

Subdivision. There are four subdivisions. Two divide nonvascular plants into the thallophytes, or algae and fungi, and the bryophytes, or the mosses and liverworts. Vascular plants are divided into the pteridophytes, consisting of the ferns and their allies, and the spermatophytes, or the seed-bearing plants. We will continue our outline with the seed-bearing plants.

Class. If it is a seed-bearing plant, is it a gymnosperm or an angiosperm? Gymnosperms include four to six orders (botanists differ on this point), of which the best known is the conifers, or cone-bearing evergreens. Conifers include pines, firs, cedars, cypresses, and araucarias. The name gymnosperm means "naked seed" and refers to the fact that the seeds do not develop within a fruiting structure such as an ovary. Gymnosperms do not have flowers.

Angiosperms have flowers, and their seeds develop within an ovary. They also differ in certain vegetative structures.

Subclass. There are two subclasses of angiosperms: the monocotyledons and the dicotyledons.

Monocots usually have flower parts (sepals, petals, stamens, and pistils) in multiples of threes, and their long, straplike leaves have parallel veins. Their roots are often bulbs, corms, or rhizomes. Familiar monocot families include the grasses, Gramineae, the lilies, Liliaceae, and the iris, Iridaceae. Plants such as orchids, Orchidaceae, sedges, Cyperaceae, arums, Araceae, and amaryllis, Amaryllidaceae, are also monocots.

Dicots have flower parts in multiples of four or five, their leaves are net-veined, and they usually have fibrous roots. Most flowering plant families are dicots.

Order. Orders are groups of related families. That is, they have certain phyletically important characteristics in common.

Family. Families are groups of related genera.

∴ WHAT'S IN A NAME? ∴

Genus. Genera are groups of related species.

Species. The species is the basic unit of classification, but oddly enough there is no universally accepted definition of what a species actually is! The best we can do is call it a kind of plant that differs in essential ways from other kinds of plants and can reproduce itself.

· F·I·F·T·E·E·N ·

Identifying Plants

Sweet Pepperbush
Clethra alnifolia

Begin with a good field guide and a hand lens. The most popular lens is a small folding one that easily slips in your pocket or hangs on a cord around your neck. It magnifies ten times (10×), which is perfect for field work. A small sharp knife or razor is helpful, too, and a ruler, preferably one that has gradations in centimeters and millimeters as well as inches, because many botanical manuals use the metric system.

USING A BOTANICAL KEY

There is no doubt that identifying an unknown plant can be daunting to the amateur. The plethora of obscure botanical terms causes many to give up before they have begun. For instance, there are over twenty terms for the shape of a leaf, nearly as many to describe its margins, and several more to describe its tip and its base. If a botanist wants to describe the hairy surface of a leaf, he or she has over sixty terms to choose from! No wonder so many give up in frustration.

Yet a botanical key is the most orderly of systems, proceeding one step at a time from the general to the specific characteristics of a plant. At each step you need choose only one of two possibilities. Your choice then leads to the next

∴ IDENTIFYING PLANTS ∴

step, where you again choose between two possibilities, and so on until you have arrived at a precise identification. Botanical manuals usually have a key for plant families at the beginning of the book; then the section for each family has a key for the genera contained within it; and each genus a key for the species.

This is the theory. In practice, the choices are not so easy. If you should pick the wrong alternative, you will immediately veer off in the wrong direction, and the plant described at the end of your journey will not bear the slightest resemblance to the one you hold in your hand! Therefore, before you start, examine the plant carefully, so you will have a good idea of its characteristics. It is also important to have all the necessary pieces. The basal leaves of wildflowers often differ from those on the stem, and in the case of woody plants, the bud, twig, and leaf may be just as important as the flower and fruit.

It is impossible to know every botanical term by heart, but many manuals and field guides have glossaries and with practice you will soon acquire a good working knowledge of those most frequently used.

As you proceed down the steps of the key, keep checking yourself. For instance, if your plant keys out to be a member of the mustard family, Cruciferae, read the description of this family to see if it fits. If you come to a place where both answers seem equally apt, mark it and proceed down each way in turn. When you come to a choice between two irrelevant descriptions, you will know you have taken the wrong road and can then go back and take the other branch. You may have to repeat this maneuver several times in a long key. Never try to go backward; it simply won't work. Also, always check your final answer against the written description of the species. Once you have a good working knowledge of twenty or thirty of the most common families, this task will be considerably easier.

If after all this the key still doesn't work, take heart: it may not be your fault. Keys vary in quality and usefulness. The author of the key might give ambiguous choices or fail to include all the alternatives, a fact particularly true of keys that try to cover large numbers of plants. They must concentrate on the average plant and cannot include every rare species or abnormal individual. Remember that large genera such as asters, *Aster*, and goldenrod, *Solidago*, have many species that hardly differ from one another. Others, such as hawthorn, *Cretaegus*, or shadbush, *Amelanchier*, tend to hybridize freely, often producing individuals that don't fit the textbook descriptions.

Amateurs tend to rely heavily on pictures, but they too can vary widely. Drawings are almost always better than photographs because the salient characteristics of the plant can be easily emphasized, but remember that "looking like the picture" is not enough for the professional. The species must conform to the written description before the identification can be considered accurate.

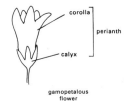

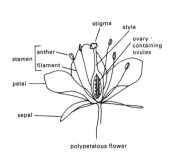

PARTS OF A FLOWER

Parts of a Flower

Plants are sorted into families and genera according to the structure of their flowers, so a working knowledge of flower parts will make identification considerably easier.

If you look closely at a flower such as a single rose or apple blossom, you will find that it has four rings of parts. The outer ring, called the *calyx*, is made up of the *sepals*. The next is the *corolla*, made up of the *petals*. Together, the calyx and the corolla form the *perianth*. If each petal and sepal is separate from the next, as in the rose, the flower is said to be *polypetalous* and *polysepalous*. If they are joined, as in a morning glory, the flower is *gamopetalous* or *gamosepalous*.

The next ring within the perianth is the *androeceum*,

literally the "male household." It consists of *stamens* (the male organs), each with a stalk called a *filament* and a pollen sac called an *anther*. The *gynoecium*, or "female household," is the fourth ring, found in the very center of the flower. The egg-containing organs, called *ovules*, are housed within an *ovary*. At the top of the ovary is a narrow neck called a *style* that is topped by the *stigma*. The flower becomes fertilized when pollen grains fall on the stigma, and the ovules then develop into seeds.

When all four of these rings are present, the flower is said to be "complete." It is "incomplete" if one or more rings are missing. A "perfect" flower has both male and female parts. An "imperfect" flower is all male or all female. If neither sex organ is present, the flower is sterile. All these parts come in a variety of shapes and sizes, sometimes imitating each other so closely that it is hard to tell a sepal or a stamen from a petal. However, if you cut a flower that is missing either sepals or petals in cross section, you can still see that two rings make up the perianth.

When you look down from above on a flower such as a single rose, you can see that a line drawn from any one point to any other point would bisect the flower into two identical halves. In this case the flower is said to be radially symmetrical, or *actinomorphic*. If a line drawn from top to bottom would result in equivalent halves, but a line drawn from side to side would not, the flower is bilaterally symmetrical, or *zygomorphic*. The spotted jewelweed, *Impatiens capensis*, is a zygomorphic flower.

The way the flowers are arranged on the stem is known as the *inflorescence*. The flowering stem is called the *peduncle*. A peduncle with stemmed flowers arranged singly along it is called a *raceme*. A lobelia is a raceme. If the flowers arise directly from the peduncle without individual stems of their own, they are said to be *sessile* and the inflorescence is a *spike*. A *panicle* is a raceme with two or more flowers on each branch.

∴ FIELD AND FOREST ∴

An *umbel* is a cluster of stalked flowers radiating from a single point. Queen Anne's lace, *Daucus carota*, is a compound umbel. A *head* is a dense cluster of tiny flowers called *florets*. In composites, there are two kinds of florets that make up the head: the disk florets in the center and the ray florets that surround it like petals.

LEAVES

alternate

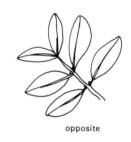

opposite

ARRANGEMENT OF
LEAVES ON A STEM

As mentioned in the beginning of this chapter, there are innumerable terms describing the shape of a leaf, as well as its margin, its tip, and its base. Still, it is helpful to know that some leaves are *simple*, like an elm, and some *compound*, like a hickory. The leaf of a hickory is *pinnately compound*, which means the individual leaflets are lined up along a central axis, as opposed to a buckeye leaf which is *palmately compound*: the leaflets arise from the base like the fingers of a hand. A leaf may also be *pinnately lobed*, like an oak, or *palmately lobed*, like a maple. Or it may be pinnately or palmately veined.

Woody plants are said to be *alternate* when the leaves and branches alternate along the stem or *opposite* when they are arranged in pairs. It is important to notice this arrangement when identifying trees and shrubs. Most woody plants are alternate. All those that are opposite are covered by the phrase MAD CAP, a handy assist for the memory that stands for Maple, Ash, Dogwood, and shrubs of the Caprifoliaceae, or honeysuckle, family.

· APPENDIX A ·

Resources

NATIVE PLANT NURSERIES

A comprehensive list of sources of native plants, both woody and herbaceous, is published by the New England Wild Flower Society, Inc., Garden in the Woods, Hemenway Road, Framingham, MA 01701. In compiling this list the Society surveyed nearly 500 North American nurseries and included only those that propagate at least 80 percent and wild collect no more than 5 percent of their native sales inventory. Every native plant gardener should have a copy.

TOOLS

The Weed Wrench is a useful tool for pulling up woody plants. It is available from New Tribe, 3435 Army Street #330, San Francisco, CA 94110. Tel: (415) 647-0430

APPENDIX B

Organizations

National Wildflower Research Center
2600 FM 973 North
Austin, TX 78725

Operation Wildflower
National Council of State Garden Clubs
Box 860
Pocasset, MA 02559

The Center for Plant Conservation
The Arnold Arboretum
Harvard University
The Arborway
Jamaica Plain, MA 02130

Society for Ecological Restoration
Madison Arboretum
University of Wisconsin
1207 Seminole Highway
Madison, WI 53711

NORTHEAST

District of Columbia
Botanical Society of Washington
Department of Biology—NHB/166
Smithsonian Institution
Washington, DC 20560

APPENDIX B

Maine
Josselyn Botanical Society
% Dr. Charles D. Richards, President
Deering Hall
University of Maine
Orono, ME 04469

Maryland
Chesapeake Audubon Society
Rare Plant Committee
P.O. Box 3173
Baltimore, MD 21228

Massachusetts
New England Botanical Club
Botanical Museums Ave.
Oxford Street
Cambridge, MA 02138

New England Wild Flower Society, Inc.
Garden in the Woods
Hemenway Rd.
Framingham, MA 01701

New Jersey
New Jersey Native Plant Society
% Frelinghuysen Arboretum
Box 1295R
Morristown, NJ 07960

New York
New York Flora Association
% New York State Museum
3132 CEC,
Albany, NY 12230

Torrey Botanical Club
New York Botanical Garden
Bronx, NY 10458

∴ ORGANIZATIONS ∴

Pennsylvania
Bowman's Hill State Wildflower Preserve
Box 345
Pineville, PA 18946

Pennsylvania Native Plant Society
1806 Commonwealth Building
316 Fourth Ave.
Pittsburgh, PA 15222

Philadelphia Botanical Club
Academy of Natural Sciences
19th and the Parkway
Philadelphia, PA 19103

SOUTH

Florida
Florida Native Plant Society
1203 Orange Ave.
Winter Park, FL 32789
(Chapters around the state)

Louisiana
Louisiana Native Plant Society
% Richard Johnson
Route 1, Box 151
Saline, LA 71070
(Chapters around the state)

North Carolina
North Carolina Wildflower Preservation Society
% North Carolina Botanical Garden
UNC-CH Totten Center 457-A
Chapel Hill, NC 27514

South Carolina
Southern Appalachian Botanical Club
Department of Biological Sciences
University of South Carolina
Columbia, SC 29208

APPENDIX B

MIDWEST

Illinois
Illinois Native Plant Society
Department of Botany
Southern Illinois University
Carbondale, IL 62901

Natural Areas Association
320 South Third St.
Rockford, IL 61108

Indiana
A.N.V.I.L. (Assoc. for use of Native Vegetation in
    Landscape)
871 Shawnee Ave.
Lafayette, IN 47905

Kansas
Kansas Wildflower Society
Mulvane Art Center
Washburn University
17th and Jewell St.
Topeka, KS 66621

Grassland Heritage Foundation
5450 Buena Vista
Shawnee Mission, KS 66205

Save the Tallgrass Prairie, Inc.
4101 West 54th Terrace
Shawnee Mission, KS 66025

Michigan
Michigan Botanical Club
Matthaei Botanical Gardens
1800 Dixboror Road
Ann Arbor, MI 48105

∴ ORGANIZATIONS ∴

SOUTHWEST

Arizona
Arizona Native Plant Society
P.O. Box 41206 Sun Station
Tucson, AZ 85717

New Mexico
Native Plant Society of New Mexico
P.O. Box 5917
Santa Fe, NM 87502

Texas
Native Plant Society of Texas
P.O. Box 23836-TWU Station
Denton, TX 76204

WEST

Utah
Utah Native Plant Society
% The State Arboretum
University of Utah, Building 436
Salt Lake City, UT 84112

Wyoming
Wyoming Native Plant Society
P.O. Box 1471
Cheyenne, WY 82001

FAR WEST

California
California Native Plant Society
909 12th Street, Suite 116
Sacramento, CA 95814
(Chapters around the state)

∴ APPENDIX B ∴

Oregon
Native Plant Society of Oregon
℅ Dr. Frank A. Lang
Department of Biology
Southern Oregon State College
Ashland, OR 97520

Washington
Washington Native Plant Society
℅ Dr. Arthur Kruckeburg
Department of Botany
University of Washington
Seattle, WA 98195

CANADA

The Canadian Wildflower Society
35 Bauer Crescent
Unionville, Ontario L3R 4H3
Canada

· BIBLIOGRAPHY ·

IDENTIFICATION

Brown, Lauren. *Grasses, an Identification Guide.* Boston: Houghton Mifflin, 1979. 240 pp.

Elias, Thomas S. *The Complete Trees of North America.* New York: Van Nostrand Reinhold, 1980. 948 pp.

Gleason, Henry A., and Arthur Cronquist. *Manual of Vascular Plants of Northeastern United States and Adjacent Canada.* New York: Van Nostrand, 1963. 810 pp.

Lellinger, David B. *A Field Manual of the Ferns and Fern-Allies of the U.S. and Canada.* Washington, D.C.: Smithsonian Institution Press, 1985. 389 pp.

Newcomb, Lawrence. *Newcomb's Wildflower Guide.* Boston: Little Brown, 1977. 463 pp.

Peterson, Roger Tory, and Margaret McKenny. *A Field Guide to Wildflowers of Northeastern and North-Central North America.* Boston: Houghton Mifflin, 1968. 420 pp.

Petrides, George A. *A Field Guide to Trees and Shrubs.* Boston: Houghton Mifflin, 1958. 431 pp.

Radford, Albert, E. Ahles, Harry E. Bell, C. Ritchie. *Manual of the Vascular Flora of the Carolinas.* Chapel Hill: University of North Carolina Press, 1968. 1,183 pp.

Silberhorn, Gene M. *Common Plants of the Mid-Atlantic Coast.* Baltimore: Johns Hopkins University Press, 1982. 256 pp.

Stokes, Donald W. *The Natural History of Wild Shrubs and Vines.* Chester, Conn.: Globe Pequot Press, 1989.

GARDENING

Art, Henry. *A Garden of Wildflowers: One Hundred One Native Species and How to Grow Them.* Pownal, Vt.: Storey Communications, 1986. 304 pp.

———. *The Wildflower Gardener's Guide: Northeast, Mid-Atlantic, Lake States, and Eastern Canada Edition.* Pownal, Vt.: 1987. 192 pp.

Barry, John M. *Natural Vegetation of South Carolina.* Columbia, S.C.: University of South Carolina Press, 1980. 214 pp.

Bir, Richard E. *Growing and Propagating Showy Native Woody Plants.* Chapel Hill: University of North Carolina Press, 1991. 230 pp.

Blumer, Karen. *Long Island Native Plants for Landscaping: A Source Book; Sources on Long Island for Environmentally Sound and Beautiful Landscaping Alternatives.* Brookhaven, N.Y.: Growing Wild Publications, 1990. 154 pp.

Bruce, Hal. *How to Grow Wildflowers and Wild Shrubs and Trees in Your Own Garden.* New York: Van Nostrand Reinhold, 1982. 299 pp.

Brumback, William E., and David Longland. *Garden in the Woods Cultivation Guide.* Framingham, Mass.: New England Wildflower Society, 1986. 51 pp.

Diekelmann, John, and Robert Schuster. *Natural Landscaping: Designing with Native Plant Communities.* New York: McGraw-Hill, 1982. 276 pp.

Dirr, Michael A., and Charles W. Heuser, Jr. *The Reference Manual of Woody Plant Propagation: From Seed to Tissue Culture.* Athens, Ga.: Varsity Press, 1987. 239 pp.

Foote, Leonard E., and Samuel B. Jones, Jr. *Native Shrubs and Woody Vines of the Southeast: Landscaping Uses and Identification.* Portland, Oreg.: Timber Press, 1989. 260 pp.

Foster, G. Gordon. *Ferns to Know and Grow.* Portland, Oreg.: Timber Press, 1984. 227 pp.

∴ BIBLIOGRAPHY ∴

Gottehrer, Dean. *Natural Landscaping*. New York: McGraw-Hill, 1982. 182 pp.

Henderson, Carrol L. *Landscaping for Wildlife*. St. Paul: Minnesota Department of Natural Resources, 1987. 145 pp.

Horton, James H., ed. *A Source Book of Information on Horticulturally Useful Native or Naturalized Plants of the Southeastern United States*. Cullowhee, N.C.: Western Carolina University, 1985. 87 pp.

Kenfield, Warren G. *The Wild Gardener in the Wild Landscape*. New London, Conn.: Connecticut Arboretum, Connecticut College, 1990. Reprint. (Originally published in 1966.) 232 pp.

Martin, Laura C. *The Wildflower Meadow Book: A Gardener's Guide*. Charlotte, N.C.: East Woods Press, 1986. 303 pp.

Mooberry, F. M., and Jane H. Scott. *Grow Native Shrubs in Your Garden*. Chadds Ford, Pa.: Brandywine Conservancy, 1980. 68 pp.

Moreton, Marion B. *Ways with Wildflowers: A Guide to Native Plant Communities*. Washington's Crossing, Pa.: Bowman's Hill Wildflower Preserve Association, 1983. 67 pp.

New England Wildflower Society. *Meadows and Meadow Gardening*. Wildflower Notes, vol. 5(1). Framingham, Mass., 1990. 43 pp.

Penn, Cordelia. *Landscaping with Native Plants*. Winston-Salem, N.C.: John F. Blair, 1982. 226 pp.

Phillips, Harry. *Growing and Propagating Wild Flowers*. Chapel Hill, N.C.: University of North Carolina Press, 1985. 330 pp.

Sawyers, Claire, ed. *Gardening with Wildflowers and Native Plants*. Plants and Gardens, vol. 45(1). Brooklyn, N.Y.: Brooklyn Botanic Garden, 1990. 96 pp.

Smyser, Carol A. *Nature's Design: A Practical Guide to Natural Landscaping*. Emmaus, Pa.: Rodale Press, 1982. 390 pp.

· INDEX OF PLANT NAMES ·

Abies balsamea, 30
Abies fraseri, 31
Acer
   A. negundo, 24
   A. platanoides, 20, 113
   A. pseudoplatanus, 20, 137
   A. rubrum, 39, 74
   A. saccharinum, 24, 74
   A. saccharum, 24, 39
Achillea millefolium, 54, 63
Acorus calamus, 75
Actinomorphic flowers, 171
Aesculus hippocastanum, 38
Aesculus octandra, 38
Agalinus maritima, 84
Agrostis, 161
Alder, 74
Alfalfa, 57, 71
Algae, 33, 165
Alien plants, 19–22, 24, 68
   damage from, 114
   effect on landscape, 134, 135
   elimination of, 144
   in open fields, 142
   in woodlands, 137, 139, 140
Alliaria officinalis, 21, 68, 137, 139
Alnus rugosa, 74
Amaryllidaceae/Amaryllis, 166
Ambrosia, spp., 19

Ammophila breviligulata, 95
Amorpha canescens, 59
Ampelopsis brevipedunculata, 20
Amphicarpa bracteata, 62
Anaphalis margaritacea, 66
Androeceum, 170–71
Andromeda
   Japanese, 47
   A. glaucophylla, 92
Andropogon, 143
   A. elliotii, 22
   A. gerardi, 59
   A. virginicus, 22, 58, 66, 143
Anemone, 42
   A. quinquefolia, 42
Anemonella thalictroides, 42
Angiosperms, 164, 166
Annuals, 9, 18–19, 29, 162
Antennaria, 66
Anther, 171
Apios americana, 62
Aquatic plants, 72–73, 77, 78
Aquilegia canadensis, 47–48
Araceae, 75, 166
Aralia hispida, 48
Aralia spinosa, 50
Araucarias, 166
Arctostaphylos uva-ursi, 99, 106, 123
Arenaria peploides, 96

Arenaria stricta, 109
Arethusa bulbosa, 90
Arisaema atrorubens, 43–44, 75
Aristida dichotoma, 66
Arrow Arum, 79
Arrowhead, 76, 79
Arrowwood, 99
Artemisia stelleriana, 97
Artemisia tridentata, 28
Arthrophyta, 33
Arum family, 75, 166
Asclepias
   A. incarnata, 69
   A. lanceolata, 84
   A. syriaca, 63
   A. tuberosa, 60, 66
   A. verticillata, 109
Ash, 5, 172
   white, 22, 39, 44, 48, 139
Aspen, 31
Aster(s), 19, 55, 60, 64, 65, 67, 99, 143, 170
   calico, 65
   New England, 65
   purple-stemmed, 69
   salt marsh, 84
   serpentine, 109
   showy, 67
   silvery, 107
   stiff, 67, 99
   white, 65

INDEX

Asteraceae, 55, 153
  A. concolor, 107
  A. depauperatus, 109
  A. ericoides, 65
  Aster family, 62–63, 153
  A. lateriflorus, 65
  A. linariifolius, 67, 99
  A. novae-angliae, 65
  A. oblongifolius, 110
  A. pilosus, 65
  A. puniceus, 69
  A. simplex, 65
  A. spectabilis, 67, 107
  A. tenuifolius, 84
Austral plants, 51
Autumn olive, 21
Azalea(s), 153
  coast, 50
  swamp, 50, 74

Baccharis halimifolia, 83
Bald cypress, 86
Baptisia tinctoria, 60, 99, 146
Bayberry, 50, 83, 98, 104, 123
Beach plants, 96
Beach-heather, 97–98, 106
Bean, wild, 62
Bearberry, 99, 106, 123
Beech, 23, 24, 39, 48, 136, 161
  family, 45
Bellwort, 151
Bergamot, lavender, 66
Betula, 24, 139
  B. lutea, 48
  B. nigra, 146
  B. papyrifera, 30
Betulaceae, 40
Bidens, 70–71
  B. coronata, 80
  B. laevis, 80
  B. polylepis, 70–71
Biennials, 9, 18, 55, 143

Birch, 24, 139
  family, 40
  paper, 30
  river, 146
  yellow, 48
Bird's foot trefoil, 61
Bitter cress, Pennsylvania, 68
Bitternut, 46
Bittersweet, 19, 20, 136, 139, 140
  Oriental, 19, 20, 116, 138, 143
Black-eyed Susan, 19, 60, 64
Blackberry, 21, 23, 140, 143
Blackspot, 130
Bladderwort, 89
Blazing Star, 66
Bloodroot, 55, 161
Blue-eyed grass, 64
Bluebell, Virginia, 68
Bluebells, 158
Blueberry, 47, 92, 93, 99, 102, 105, 107, 123, 140, 143
  low-bush, 106–7
Bog Asphodel, 91–92
Bog bean, 90–91
Bog bunchflowers, 91
Bog hyacinth, 91
Bog plants, 94, 100
Bog rosemary, 92–93
Boneset, 66
Boreal forest, plants, 31, 51, 92, 104
Botanical nomenclature, 158–62
Bouncing Bet, 54
Box elder, 24
Bracken, 106
Brassica, 19, 63
Brazilian pepper, 20
Bromeliaceae, 86
Bromeliad family, 86

Broom crowberry, 104
Broomsedge, 22, 58, 66, 123, 142, 143
Bryophytes, 87, 165
Buchloe dactyloides, 59
Buckeye
  leaf, 172
  sweet, 38
Buckwheat family, 21
Bull bay, 50
Bulrushes, 76
Bunchberry, 48
Bunchlily family, 107
Bush fallow, 53
Bush honeysuckle, 20, 138, 139, 145, 158
Butter and eggs, 54, 64
Buttercup, 54, 57
Butterfly weed, 60, 66
Buttonbush, 74

Cacti, 29
Cakile edentula, 96
Calamagrostis canadensis, 76
Calopogon tuberosus, 90
Caltha palustris, 74
Calyx, 170
Camassa scilloides, 151
Campsis radicans, 140
Canopy trees, 39, 44, 122–23, 136, 139
Caprifolaceae, 172
Cardamine bulbosa, 68
Cardamine pensylvanica, 68
Cardinal flower, 69
Carduus nutans, 55
Carnation family, 64
Carnivorous plants, 88–89
Carpinus caroliniana, 40
Carya, spp. 23, 43, 46
  C. cordiformis, 46
  C. glabra, 46

INDEX

C. ovata, 46
C. tomentosa, 46
Caryophyllaceae, 64, 96, 107
Castanea dentata, 43
Catalpa, 102
Catchfly, pink, 107
Cattail, 73, 75, 79
Cattail, broad-leaf, 75
Ceanothus americanus, 46
Cedar, 166
    red, 22, 122, 123, 139, 142, 143, 158
Celandine, 54
Celastrus orbiculatus, 19
Celastrus scandens, 19, 20
Centaurea, spp., 54
Centaurea cineraria, 97
Cephalanthus occidentalis, 74
Cerastium arvense, var. *villosum*, 109
Cercis canadensis, 40
Chamaecyparis thyoides, 93
Chamaedaphne calyculata, 92
Chelidonium majus, 54
Chelone glabra, 69
Chenopodiaceae, 83, 97
    C. album, 64
    C. hybridum, 64f
Cherry, 22
    black, 39, 139, 146
Chestnut, American, 43
Chickweed, 109
Chicory, 64
Chimaphila maculata, 146
Chrysanthemum leucanthemum, 54
Chrysanthemum parthenium, 54
Chrysopsis mariana, 107
Cichorium intybus, 64
Cirsium, 55
    C. altissimum, 55
    C. arvense, 55

C. discolor, 55
C. pumilum, 55
C. vulgare, 55
Cistaceae, 98
Cladrastis lutea, 12
Class, 166
Classification, 162–67
Claudium jamaicensis, 76
Claytonia virginica, 14, 42
Clethra acuminata, 40
Clethra alnifolia, 74
Cliff green, 110
Clotbur, beach, 96
Clover
    alsike, 61
    Kate's mountain, 110
    purple prairie, 59
    red, 61
    white, 61
    yellow sweet, 61
Club moss, 4, 51
Cnaphalium obtusifolium, 66
Cocos nucifera, 26
Columbine, wild, 47–48
Compositae, 55, 62–63, 80, 172
Comptonia peregrina, 106, 146
Coneflower, green-headed, 68
Conifers, 5, 10, 31, 49–50, 105, 166
Convolvulus purshianus, 110
Coptis groenlandica, 30
Corema conradii, 104
Coreopsis, yellow (coreopsis lauceolata), 66
Cornus
    alternifolia, 48
    C. amomum, 68
    C. canadensis, 48
    C. florida, 10, 22, 40
Corolla, 170
Coronilla varia, 61

Cotton grass, 90
Crabgrass, 18
Cranberry, 92–93
Crataegus, spp., 22, 170
    C. viridis, 124
Cress, spring, 68
Crown vetch, 61
Cruciferae, 68, 96
Cudweed, 66
Culms, 56
Cynodon dactylon, 58
Cyperaceae, 166
Cyperus strigosus, 76
Cypress, 166
Cypress spurge, 158
Cypripedium, spp., 90

Dactylis glomerata, 58
Daisy, field, 54
Daisy family, 55, 62
Daisy fleabane, 19
Dandelion, 19, 54, 63
Dangleberry, 107
Datura stramonium, 96
Daucus carota, 18, 54, 64, 172
Daylily, tawny, 54–55
Deciduous plants, 10
Decodon verticillatus, 75
Dentaria, spp., 42
Desmodium, spp., 62
Dianthus armeria, 64
Dicentra cucullaria, 42
Dicotyledons, 62, 128, 150, 163, 166
Digitaria sanguinalis, 18
Dionaea muscipula, 89
Diospyros virginiana, 51
Dipsacus sylvestris, 54
Distichlis spicata, 83
Division, 165
Dogwood
    alternate leaved, 48

INDEX

Dogwood (*cont'd*)
   flowering, 10, 22, 40, 46, 48, 122, 123, 139, 143, 146, 172
   gray-stemmed, 68
Drosera, 88
Drosera filiformis, 88
Duckweed, 78
Dune plants, 96–99
Dutchman's breeches, 42

Eichhornia crassipes, 78
Elaegnus umbellata, 21
Elderberry, 40, 48, 68, 74
Eleocharis, spp., 76
Elliot's beardgrass, 22
Elm, 120, 152, 172
Epigaea repens, 106, 153
Epiphytic plants, 86
Equisetum arvense, 33
Eragrostis, spp., 58
   E. spectabilis, 58
Ericaceae, 47, 92, 93, 99, 106–7, 153
Erigeron annuus, 19, 63
Eriocaulon septangulare, 91
Eriophorum virginicum, 90
Erythronium americanum, 42, 151
Eupatorium, spp., 153
   E. aromaticum, 109
   E. fistulosum, 69
   E. hyssopifolium, 66
   E. rugosum, 47
Euphorbia corollata, 67
Euphorbia polygonifolia, 98
Evening primrose, common, 60, 64
Evening primrose family, 153
Everlastings, 66
Exotic plants, 113, 114, 134, 143, 144–47

Fagaceae, 45
Fagus grandifolia, 23, 39, 48
Fame flower, 109
Fenrose, 84
Fern(s), 4, 39, 48, 136, 165
   Christmas, 41
   cinnamon, 41
   in garden, 123
   interrupted, 41
   in mixed mesophytic woods, 41
   New York, 41
   in plant succession, 23
   risk to, 151
Fescue, 58
Festuca elatior, 58
Fetterbush, 158
Feverfew, 54
Filament, 171
Fir, 166
   balsam, 30
   Fraser, 31
Flag, blue, 75
Flax, 54
Fleabane, 63
   salt marsh, 84
Floating heart, 77
Floating-leaved plants, 77
Floodplain species, 23, 74, 120
Floret, 172
Flowering plants, 5, 33, 103–4, 162–63, 166
Flowers
   of dry (xeric) fields, 65–67
   of mesic fields, 54, 55, 62–65, 141
   prairie, 60, 63, 66
   of salt marsh meadow, 84–85
   structure of, 162–63, 170, 171, 172

   of wet (hydric) meadow, 67–71
Forage plants, 56–61
Forest, Coastal Plain, 49–50
Franklinia, 41
F. alatamaha, 14–15, 161
Fraxinus americana, 22, 39, 48
Free-floating aquatics, 78
Frostweed, 65

Galactia regularis, 62
Gamopetalous flowers, 170
Gamosepalous flowers, 170
Gaultheria procumbens, 47
Gaylussacia, 47, 99, 107
   G. baccata, 106–7
   G. frondosa, 107
Gentian
   bottle, 71
   closed, 71
   family, 60, 71, 77
   fringed, 71
   Pine Barrens, 107–8
Gentiana, 60
   G. andrewsii, 71
   G. autumnalis, 107
   G. clausa, 71
   G. crinita, 71
Gentianaceae, 77
Geranium, wild, 55
Gerardia, seaside, 84
Ginseng, 8
Glaucium flavum, 97
Goatsbeard, 66
Golden Alexander, 67–68
Golden Aster, Maryland, 107
Golden Aster, sickle-leaved, 98, 121
Golden club, 75
Goldenrod, 19, 22, 60, 64–65, 99, 107, 146, 153, 170
   Canada, 65

∴ INDEX ∴

early, 64
Elliot's, 84–85
field, 143, 144
lance-leaved, 69
rough-stemmed, 64–65
salt marsh, 84–85
seaside, 84, 96
tall, 64–65
Goldenseal, 8
Goldthread, 30
Goosefoot, 64
Goosefoot family, 83, 97
Graminae, 56, 80, 166
Grape, wild, 41
Grass, 22, 28, 55, 56–60, 70, 123, 134, 146, 147, 153
Bermuda, 58
big bluestem, 59, 60
blue joint, 76
in bogs, 86
buffalo, 59–60
dune, 95–96, 97
family, 56, 80, 166
flowers of, 56
imported forage, 56–58, 66, 141, 143
Indian, 59, 143
marsh, 76–77, 82–83
Kentucky blue, 57
little bluestem, 22, 59, 143
love, 58
meadow fescue, 58
orchard, 58
poverty, 66
purple love, 58
purple top, 58, 123
reed canary, 76–77
saltwater cord-, 82
spike, 83
switch, 58
Grass pink, 90
Grasses, prairie, 59–60

Green lacewing, 129–30
Greenbrier, 41, 140
Ground pine, 33
Ground-cherry, 66–67
Groundnuts, 62
Gum, black, 40, 50
Gum, sweet, 39, 50, 51, 120
Gymnosperms, 49, 86, 166
Gynoecium, 171

Halesia carolina, 38
Halesia monticola, 38
Halophytes, 97
Hamamelis virginiana, 7, 22, 40
Hardhack, 69
Hardwoods, 24
Hawkweed, orange, 91
Hawthorne, 22, 122, 143, 170
green, 124
Heath family, 47, 92, 93, 99, 106, 153
Heaths, 28, 104
Helenium, spp. 69, 70
Helianthus, spp. 60, 69
H. angustifolius, 80
H. annuus, 70
H. giganteus, 70
H. tuberosus, 70
Hellonias bullata, 91
Hemerocallis fulva, 54
Hemlock, 24, 39, 48, 49, 136
Hepatica, 41, 55
H. acutiloba, 41
H. americana, 41
Herbaceous plants, 10, 28, 37, 115, 127, 153
of dry oak woods, 47–48
fire and, 146
in fields, 142
in mixed mesophytic woods, 39–40
Herbals, 54, 162

Hercules' club, 50
Hibiscus moscheutos, 79
Hibiscus palustris, 79
Hickory, 23, 43, 44, 46, 48, 102, 136, 146
pignut, 46
shagbark, 46
leaves, 172
High tide bush, 83
Hobblebush, 48
Hog-peanut, 62
Holly, American, 50, 83–84
Honeysuckle, 116, 136, 138, 139, 143, 172
Japanese, 19–20, 145
Hop hornbeam, 40
Horn-poppy, 97
Hornbeam, American, 40
Horse chestnut, 38
Horse nettle, 57, 67
Horsetail, field, 33
Houstonia caerulea, 63
Huckleberry, 47, 99, 197, 140, 146
black, 106–7
Hudsonia ericoides, 106
Hudsonia tomentosa, 97–98, 106
Hyacinth, wild, 151
Hydrastis canadensis, 8
Hypericum, spp., 64

Ilex
I. glabra, 84, 94
I. opaca, 50, 83–84
I. verticillata, 50, 74, 84
Impatiens capensis, 68, 171
Impatiens pallida, 68
Indian corn, 56
Indigo, wild, 60
Inkberry, 84, 94
Iridaceae, 64, 150, 166

∴ INDEX ∴

Iris family, 55, 56, 64, 150, 166
  I. pseudacorus, 75
  I. versicolor, 75
  yellow, 75
Ironweed, 60
  New York, 69
Iva frutescens, 83

Jack-in-the-pulpit, 43–44, 75
Jerusalem artichoke, 70
Jewelwood, 68–69
  spotted, 68, 171
  yellow, 68
Jimson weed, 96
Joe Pye weed, 153
  hollowstemmed, 69
Juglans nigra, 11, 24, 39, 93
Juncaceae, 76
  J. canadensis, 76
  J. effusus, 76, 79
  J. gerardi, 83
  J. roemerianus, 83
Juniperus virginiana, 22

Kalmia, 161
  K. angustifolia, 107
  K. latifolia, 47, 107
  K. polifolia, 92, 93
Kingdom, plant, 165
Knapweeds, 54
Knotweed
  Japanese, 21
  seabeach, 98
Kosteletzkya virginica, 84
Kudzu, 21, 61–62

Labrador tea, 92, 93
Ladies' tresses, 90
Lady slippers, 90, 150–51
Lambsquarters, 64
Larix laricina, 30, 93
Lathyrus maritimus, 97

Laurel, 16, 41, 105, 107
  bog, 92, 93
  mountain, 47, 107, 152, 153
  sheep, 107
Lavender, sea, 84
Leadplant, 59
Leatherleaf, 92
Ledum groenlandicum, 92
Legumes, 55, 56, 60–62
Leguminosae, 60, 153
Leiophyllum buxifolium, 106
Lenni-Lenape, 75
Lepidophyta, 33
Liatris, spp. 66
Ligustrum sinense, 20
Ligustrum vulgare, 20
Liliaceae, 91, 166
  L. canadense, 69
  L. superbum, 64
Lily
  Canada, 69
  family, 55, 56, 64, 69, 91, 166
  family at risk, 150, 151
  Turk's cap, 64
Limonium nashii, 84
Linaria vulgaris, 54, 64
Lindera benzoin, 22, 40
Linum usitatissimum, 54
Liparis loeselii, 102
Liquidambar styraciflua, 39, 50
Liriodendron tulipifera, 7, 17, 37
Lithospermum canescens, 109
Liverworts, 165
Lobelias, 55, 171
  blue, 13, 69
  L. cardinalis, 69
  L. siphilitica, 13, 69
Locust, black, 109
Loesels twayblade, 102
Lonicera, 158
  L. japonica, 19–20, 145

  L. maacki, 20
  L. tatarica, 20
Loosestrife, 76
  fringed, 68, 76
  purple, 20, 70, 76, 114
  swamp, 75–76
Lotus, American, 77
Lotus corniculatus, 61
Lycopodium, spp. 4, 33, 51
Lysimachia
  L. ciliata, 68, 76
  L. nummularia, 63
  L. terrestris, 76
Lythraceae, 76
Lythrum salicaria, 20, 70, 76

Magnolias, 7, 37, 38–39, 43, 44, 104
  Fraser, 38–39
  M. fraseri, 38
  M. grandiflora, 50
  sweet bay, 50, 94
  M. tripetala, 38
  umbrella, 38
  M. virginiana, 50, 94
Maianthemum canadense, 48
Mallow, rose, 79–80, 84
Mallow family, 84
Malvaceae, 84
Mangrove, red, 6, 26
Maple(s), 5, 23, 122, 137–38, 146, 172
  Norway, 20, 113–14, 137–38
  red, 22, 24, 39, 44, 46, 48, 74, 103, 136, 139, 152–53
  silver, 24, 74
  sugar, 24, 39, 44, 48, 136, 139
  sycamore, 20, 137–38
Marsh elder, 83
Marsh marigold, 83
Marsilea quadrifolia, 77

INDEX

Mayapple, 42
Mayflower, Canada, 48
Meadow parsnip, 67–68
Medicago lupulina, 61
Medicago sativa, 57, 71
Medicinal plants, 8, 13, 48, 54, 69, 71, 75, 87, 158, 161
Medick, black, 61
Melaleuca quinquenervia, 21
Melilotus officinalis, 61
Menyanthes trifoliata, 90–91
Mertensia virginica, 68
Mikania scandens, 80
Milk pea, 62
Milkweed
 common, 63
 salt marsh, 84
 swamp, 69
 whorled, 109
Milkwort, racemed, 99
Mimulus ringens, 69
Mitchella repens, 14
Mockernut, 46
Monarda fistulosa, 66
Moneywort, creeping, 63
Monkey flower, 69
Monocotyledons, 56, 128, 150, 163, 166
Monoecious plants, 75
Morning-glory, shale, 110
Moss(es), 28, 39, 87, 165
 Spanish, 86
 sphagnum, 87–88, 91, 92
Moss phlox, 108
Mullein, 18
 flannel, 54
Mustard, garlic, 21, 68, 137, 139
Mustard, yellow, 63
Mychorrizae, 90, 150
Myrica cerifera, 50, 83, 94

Myrica pensylvanica, 50, 83, 98
Myrtle, sand, 106

Named cultivars, 151–52, 153
Narthecium americanum, 90–91
Navel orange, 10
Nelumbo lutea, 77
New Jersey tea, 46–47
Nonvascular plants, 165
Nuphar advena, 77
Nymphaea odorata, 77
Nymphoides aquatica, 77
Nymphoides cordata, 77
Nyssa sylvatica, 40
Nyssa sylvatica biflora, 40

Oak(s), 12, 23, 24, 44–46, 48, 122, 136, 163
 bear, 46, 103, 105
 black, 39, 44, 45–46, 103
 blackjack, 23, 44, 46, 51, 103, 105
 chestnut, 23, 44, 45, 46, 103
 and dry soils, 46–47
 early, 104
 in Pine Barrens, 103, 105
 post, 46, 103, 105
 red, 39, 44, 45–46, 48
 red (or black) group, 45–46
 scarlet, 45–46, 103
 Spanish, 46
 species, 44
 swamp chestnut, 44
 swamp white, 24, 44
 water, 44, 51
 white, 39, 44, 45, 103
 willow, 44, 51
Oconee bells, 41–42
Oenothera, 60
 biennis, 64
Onagraceae, 153
Opuntia humifusa, 66

Orache, 97
Orchidaceae, 89, 150, 166
Orchid family, 89–90, 102, 150–51, 166
Order, 166
Orontium aquaticum, 75
Osmunda cinnamomea, 41
Osmunda claytoniana, 41
Ostrya virginiana, 40
Ovary, 171
Ovules, 171
Oxalis montana, 30
Oxydendron arboreum, 47, 124

Pachistima canbyii, 110
Palm(s)/Palmae, 26
 coconut, 26
 date, 26
Panax quinquefolius, 8
Panicle, 171
Panicum virgatum, 58
Parsley family, 67
Parsnip, wild, 68
Parthenocissus quinquefolia, 41
Partridge berry, 14
Pastinaca sativa, 68
Pea, beach, 97
Pea family, 59, 60, 61, 62, 153, 163
Pearly everlasting, 66
Peduncle, 171
Peltandra virginica, 79
Pepperbush, sweet, 40, 74
Perennials, 9, 19, 143, 162
Perianth, 170, 171
Persimmon, 51
Petalostemum purpureum, 59
Petals, 170
Phalaris arundinacea, 76
Phaseolus polystachios, 62
Phleum pratense, 57

INDEX

Phlox, 60
  P. subulata, 108
Phoenix dactylifera, 26
Phragmites, 80–81, 114
  P. australis, 6, 80
Physalis, 66
Phytolacca americana, 23, 57
Picea
  P. glauca, 30
  P. mariana, 30, 93
  P. rubens, 31
Pickerelweed, 76, 79
Pieris japonica, 47
Pine(s), 12, 49, 122, 139, 166
  jack, 30, 104
  loblolly, 50, 51, 100
  lodgepole, 30
  longleaf, 50, 100
  pitch, 23, 50, 100, 103, 104, 105
  red, 104
  scrub, 50
  shortleaf, 50, 100, 103, 105
  slash, 100
  Virginia, 23
  white, 24, 48, 50, 136
Pine Barrens, 51, 85, 87, 90, 91, 93, 100–108
Pink
  clove, 158
  deptford, 64
  family, 96, 107
Pinus
  P. banksiana, 30, 104
  P. contorta, 30
  P. echinata, 50, 100, 103
  P. elliottii, 100
  P. palustris, 50, 100
  P. resinosa, 104
  P. rigida, 23, 50, 100, 103
  P. strobus, 24, 48

P. taeda, 50, 51, 100
P. virginiana, 23, 50
Pinxterbloom, 40, 160
Pipewort, 91
Pitcher plants, 86, 88
Pityopsis falcata, 98, 121
Plant(s)
  American, sent to Europe, 13–15
  classification of, 15, 162–67
  families, 160–61, 166, 169, 170
  names of, 157–67
  endangered, 101, 150–51
  field, 142, 143, 145
  identifying, 168–72
  knowledge of, 124
  meadow, 153
  pest, 113
  pond, 74, 79
  rare, 7, 8, 42, 101, 108–9, 110, 153
  rescue of, 154
  upland, 100, 135
Platanthera, 90
Platanus occidentalis, 24, 38, 50, 74
Pluchea purpurascens, 84
Plum, beach, 98–99
Plume flower, 42
Poa pratensis, 57
Podophyllum peltatum, 42
Pogonia, rose (*Pogonia ophioglossoides*), 90
Poison ivy, 41, 140
Poisonous plants, 57, 67, 107
Poke, 23, 57
Polygonatum biflorum, 42, 151
Polygala
  P. lutea, 91
  P. polygama, 99
  P. senega, 13

Polygonaceae, 21
  P. arifolium, 79
  P. cuspidatum, 21
  P. densiflorum, 79
  P. glaucum, 98
  P. hydropiperiodes, 79
  P. punctatum, 79
Polystichum acrostichoides, 41
Pontederia cordata, 76
Populus tramuloides, 31
Porcelain-berry vine, 20, 139, 140, 145
Portulaca, 109
  P. oleracea, 18
Prickly-pear, 66
Primrose family, 76
Primulaceae, 76
Privet, 138, 139
  Chinese, 20
  common, 20
Prunus, spp., 22
  P. maritima, 98–99
  P. serotina, 39, 139, 146
Pteridium aquilinum, 106
Pteridophytes, 165–66
Puccoon, hoary, 109–10
Pueraria hirsuta, 21, 61
Punk tree, 21
Purslane, 18
Pussytoes, 66
Pyxidanthera barbulata, 106
Pyxie moss, 106

Quaker ladies, 63
Queen Anne's lace, 18, 54, 64, 172
Quercus
  Q. alba, 39, 44, 45
  Q. biflora, 24
  Q. coccinea, 45
  Q. falcata, 46
  Q. ilicifolia, 46, 103, 105

∴ INDEX ∴

Q. marilandica, 23, 44, 46, 51, 103, 105
Q. michauxii, 44
Q. nigra, 44, 51
Q. phellos, 44, 51
Q. prinus, 23, 44, 45
Q. rubra, 39, 44
Q. stellata, 46, 103
Q. velutina, 39, 44, 45–46, 103

Raceme, 171
Ragweed, 19, 70
Ragwort, golden, 60, 63
Ranunculus acris, 54, 57
Raspberry, black, 140
Redbud, 40, 139
Reed, giant, 6, 80–81
Rhexia, spp., 91
Rhizophora mangle, 6, 26
Rhododendron, 7, 41
   R. atlanticum, 50
   R. periclymenoides, 40, 160
   R. viscosum, 50, 74
Rhus, spp., 21
   R. aromatica, 47
   R. copallina, 99
   R. radicans, 41
Rocket, sea, 96
Rockrose family, 98
Rosa
   R. carolina, 64, 99
   R. multiflora, 20, 64
   R. palustris, 68
   R. rugosa, 96
Rose(s), 21, 170
   multiflora, 20–21, 64, 109, 114, 116, 138, 139, 140, 143, 145
   pasture, 64, 99
   salt spray, 96–97
   swamp, 68

Rubus, spp., 21
Rudbeckia hirta, 19, 60, 64
Rudbeckia laciniata, 68
Ruderals, 18–19, 126–27, 134
Rue anemone, 42
Rushes, 71, 76
   black, 83
   Canada, 76
   chairmaker's, 76
   soft, 76, 79
   spike, 76

Sagebrush, 28
Sagittaria latifolia, 76, 79
St. Andrew's cross, 157
St. John's wort, 64
Salicornia, 83
Salix nigra, 24, 39, 50, 74
Salsola kali, 97
Salt hay, 82–83
Saltwort, 97
Sambucus canadensis, 40, 68, 74
Sandwort, 96
   rock, 109
Sanguinaria, 161
Saponaria officinalis, 54
Sarracenia purpurea, 86
Sarsaparilla, 48
Sassafras, 7, 13, 22, 122, 139, 143
   S. albidum, 13, 22
Scale, of garden, 123
Scale, insect, control of, 130
Scale trees, 33
Schinus terebinthifolius, 20
Schizachyrium scoparium, 22, 59
Scirpus, 76
S. americanus, 76
S. cyperinus, 76
Scrophulariaceae, 153
Sea blite, 97
Sedge, umbrella, 76

Sedges, 28, 71, 76, 86, 90, 146, 166
Seed-bearing plants, 166–67
Senecio aureus, 60
Sepals, 170
Sessile flowers, 171
Shadbush, 40, 139, 170
Shortia galacifolia, 41–42
Shortia uniflora, 42
Shrubs, 19–20, 37, 44, 47, 114, 162
   in bogs, 92
   Coastal Plain, 50
   field, 136
   forest, 39, 40, 48, 136, 139–40
   garden, 121–22, 123–24, 131
   getting rid of, 145
   heath family, 93
   identifying, 172
   imported, 134
   invasive, 139
   Pine Barrens, 105
   salt marsh, 83–84
   seaside, 98–99
Silene caroliniana, 107
Silverbell, 38, 44
   Carolina, 38
Sisyrinchium, spp., 64
Skunk cabbage, 41, 74, 75
Smartweed, 79
Smilacina racemosa, 42
Smilax, spp., 41, 140
Snakeroot, 47, 109
   dry-land, 109
   Seneca, 13
Snapdragon family, 153
Sneezeweed, 69, 70
Solanaceae, 57, 67
Solanum carolinense, 57, 67
Solidago
   S. altissima, 64

INDEX

Solidago (cont'd)
  S. canadensis, 65
  S. elliotii, 84
  S. graminifolia, 69
  S. juncea, 64
  S. nemoralis, 66
  S. rugosa, 64–65
  S. sempervirens, 84, 96
Solomon's seal, 42, 151, 157
Sonchus, spp., 64
Sorghastrum nutans, 59, 143
Sourwood, 47, 124
Sow thistle, 64
Spartina, 82–83
  S. alterniflora, 82
  S. cynosuroides, 82
  S. patens, 82
Spatterdock, 77
Specularia perfoliata, 63
Spermatophytes, 166
Spicebush, 22, 40
Spike, 171
Spikelet, 56
Spiraea tomentosa, 69
Spiranthes, spp., 90
Spirodela polyrhiza, 78
Spring beauty, 14, 42
Spring ephemerals, 23, 42–43
Spruce
  black, 30, 93
  red, 31
  white, 30
Spurge
  flowering, 67
  seaside, 98
Stamens, 15, 163, 171
Starflower, 30
Stigma, 171
Style, 171
Suaeda maritima, 97
Subclass, 166
Subdivision, 165–66

Sumac, 21, 122, 139, 143
  shining, 99
Sunbright, 109
Sundew, 88–89
Sunflower, 55, 60, 69–70
  common, 70
  narrow-leaved, 80
  tall, 70
Swamp candles, 76
Swamp-pink, 90
Sweet gale, 83
Sweetfern, 106, 146
Sweetflag, 75
Sycamore, 24, 39, 50, 74, 120
Symplocarpus foetidus, 41, 74

Talinum teretifolium, 109
Tamarack, 30, 93
Tanacetum vulgare, 54
Tansy, 54
Taraxacum officinale, 19, 54
Taxodiaceae, 86
Taxodium distichum, 86
Taxus canadensis, 48
Tearthumb, halberd-leaved, 79
Teasel, 54
Thallophytes, 165
Thaspium trifoliatum, 67
Thatch, 130–31
Thelypteris noveboracensis, 41
Thistles, 55, 57, 141, 144
  bull, 55
  Canada, 55, 141, 145
  field, 55
  nodding, 55
  pasture, 55
  tall, 55
Tick-trefoils, 62
Tickseed sunflowers, 69, 70, 71, 80
Tillandsia usneoides, 86
Timothy, 57–58

Tomato family, 57, 67
Toothwort, 42
Tragopogon pratensis, 66
Trailing arbutus, 106, 153
Trees, 114
  classification of, 162
  eastern woods, 37
  in fields, 141, 142, 143, 144
  fire and regeneration of, 146
  in garden, 121–22, 124
  identifying, 172
  imported, 134
Trees, north woods, 30–31
Trees in plant succession, 19
  released landscape, 136, 140–41
  understory, 22, 39, 40, 44, 136
  woodland, 139–40
Trientalis borealis, 30
Trifolium
  T. hybridum, 61
  T. pratense, 61
  T. reptens, 61
  T. virginicum, 110
Trillium, 42, 151
  T. grandiflorum, 151
Triodia flava, 58
Trout lilies, 42, 151
Trumpetvine, 140
Tsuga canadensis, 24, 39, 48
Tulip poplar, 7, 17, 22, 23, 37–38, 43, 44, 139
Turkeybeard, 107
Turkeyfoot, 59
Turtlehead, 69
Typha latifolia, 75

Umbel, 172
Umbelliferae, 67
Utricularia, spp., 89
Uvularia, 151

∴ INDEX ∴

Vaccinium, 47, 92, 99
   V. macrocarpum, 92
   V. oxycoccos, 92
   V. vacillans, 106
Vascular plants, 165–66
Venus' flytrap, 89
Venus' looking glass, 63, 157
Verbascum thapsus, 18, 54
Vernonia, 60
   V. noveboracensis, 69
Viburnum, 21, 123, 140
   V. acerifolium, 22, 40
   V. alnifolium, 48
   V. dentata, 99
   V. mapleleaf, 22, 40
   V. recognitum, 99
Vines, 21, 145
   alien, 19–20, 134, 142
   in forest, 40–41, 136–37, 138, 139, 140
Viola papilionacea, 63
Viola pedata, 66

Violets, 55, 63
   bird's-foot, 66
   blue, 63
Virginia creeper, 41
Vitis, spp., 41

Walnut, black, 11, 24, 39, 93
Walnut family, 46
Water hyacinth, 78
Water shamrock, 77
Water willow, 75–76
Waterlily, 77
Wax myrtle, 50, 94
Weeds, 8, 18, 53, 64
   exotic, 143–44, 146
   in garden, 126–28, 131
   in natural areas, 133–34
   in woodland, 140
White cedar, Atlantic, 93
Wild populations, 149–50
Wild rice, 56, 79
Wildflowers. See *Flowers*

Willow, black, 24, 39, 50, 74
Willows, 163
Winterberry, 50, 74, 84
Wintergreen, 47
   spotted, 146
Witch hazel, 7, 22, 40, 48
Wood sorrel, 30
Woody plants, 94, 114, 143, 153, 169, 172
Wool grass, 76
Wormwood, beach, 97

Xanthium echinatum, 96
Xerophyllum asphodeloides, 107

Yarrow, 54, 63
Yellowwood, 12
Yew, 48

Zizania aquatica, 56, 79
Zizia aurea, 67
Zygomorphic flowers, 171